AF294963

Tucholsky Wagner Zola Scott Sydow Freud Schlegel
Turgenev Wallace Fonatne
Twain Walther von der Vogelweide Fouqué Friedrich II. von Preußen
Weber Freiligrath Frey
Kant Ernst
Fechner Weiße Rose von Fallersleben Richthofen Frommel
Fichte Hölderlin
Engels Fielding Eichendorff Tacitus Dumas
Fehrs Faber Flaubert Eliasberg Ebner Eschenbach
Feuerbach Maximilian I. von Habsburg Fock Zweig
Ewald Eliot Vergil
Goethe London
Mendelssohn Balzac Shakespeare Elisabeth von Österreich
Dostojewski Ganghofer
Trackl Stevenson Lichtenberg Rathenau Doyle Gjellerup
Mommsen Tolstoi Hambruch Droste-Hülshoff
Thoma Lenz Hanrieder
Dach Verne von Arnim Hägele Hauff Humboldt
Reuter Rousseau Hagen Hauptmann Gautier
Karrillon Garschin Defoe Baudelaire
Damaschke Descartes Hebbel
Wolfram von Eschenbach Hegel Kussmaul Herder
Darwin Dickens Schopenhauer Rilke George
Bronner Melville Grimm Jerome Bebel Proust
Campe Horváth Aristoteles Voltaire Federer Herodot
Bismarck Vigny Barlach Heine
Gengenbach
Storm Casanova Tersteegen Grillparzer Georgy
Chamberlain Lessing Langbein Gilm Gryphius
Brentano Lafontaine
Strachwitz Claudius Schiller Kralik Iffland Sokrates
Katharina II. von Rußland Bellamy Schilling
Gerstäcker Raabe Gibbon Tschechow
Löns Hesse Hoffmann Gogol Wilde Vulpius
Luther Heym Hofmannsthal Gleim
Roth Klee Hölty Morgenstern
Heyse Klopstock Kleist Goedicke
Luxemburg Puschkin Homer Mörike
Machiavelli La Roche Horaz Musil
Navarra Aurel Musset Kierkegaard Kraft Kraus
Nestroy Marie de France Lamprecht Kind Kirchhoff Hugo Moltke
Laotse Ipsen Liebknecht
Nietzsche Nansen Ringelnatz
Marx Lassalle Gorki Klett
von Ossietzky May Leibniz
vom Stein Lawrence Irving
Petalozzi Knigge
Platon Pückler Michelangelo Kafka
Sachs Poe Liebermann Kock Korolenko
de Sade Praetorius Mistral Zetkin

The publishing house tredition has created the series **TREDITION CLASSICS**. It contains classical literature works from over two thousand years. Most of these titles have been out of print and off the bookstore shelves for decades.

The book series is intended to preserve the cultural legacy and to promote the timeless works of classical literature. As a reader of a **TREDITION CLASSICS** book, the reader supports the mission to save many of the amazing works of world literature from oblivion.

The symbol of **TREDITION CLASSICS** is Johannes Gutenberg (1400 – 1468), the inventor of movable type printing.

With the series, tredition intends to make thousands of international literature classics available in printed format again – worldwide.

All books are available at book retailers worldwide in paperback and in hardcover. For more information please visit: www.tredition.com

tredition was established in 2006 by Sandra Latusseck and Soenke Schulz. Based in Hamburg, Germany, tredition offers publishing solutions to authors and publishing houses, combined with worldwide distribution of printed and digital book content. tredition is uniquely positioned to enable authors and publishing houses to create books on their own terms and without conventional manufacturing risks.

For more information please visit: www.tredition.com

Letters of a Radio-Engineer to His Son

John Mills

Imprint

This book is part of the TREDITION CLASSICS series.

Author: John Mills
Cover design: toepferschumann, Berlin (Germany)

Publisher: tredition GmbH, Hamburg (Germany)
ISBN: 978-3-8491-5387-8

www.tredition.com
www.tredition.de

Copyright:
The content of this book is sourced from the public domain.

The intention of the TREDITION CLASSICS series is to make world literature in the public domain available in printed format. Literary enthusiasts and organizations worldwide have scanned and digitally edited the original texts. tredition has subsequently formatted and redesigned the content into a modern reading layout. Therefore, we cannot guarantee the exact reproduction of the original format of a particular historic edition. Please also note that no modifications have been made to the spelling, therefore it may differ from the orthography used today.

Pl. I.–One of the Lines of Towers at Radio Central
(Courtesy of Radio Corporation of America).

TO

J. M., Jr.

CONTENTS

LIST OF PLATES

LETTERS OF A RADIO-ENGINEER TO HIS SON

3 LETTER 1
ELECTRICITY AND MATTER

My Dear Son:

You are interested in radio-telephony and want me to explain it to you. I'll do so in the shortest and easiest way which I can devise. The explanation will be the simplest which I can give and still make it possible for you to build and operate your own set and to understand the operation of the large commercial sets to which you will listen.

I'll write you a series of letters which will contain only what is important in the radio of to-day and those ideas which seem necessary if you are to follow the rapid advances which radio is making. Some of the letters you will find to require a second reading and study. In the case of a few you might postpone a second reading until you have finished those which interest you most. I'll mark the letters to omit in this way.

All the letters will be written just as I would talk to you, for I shall draw little sketches as I go along. One of them will tell you how to experiment for yourself. This will be the most interesting of all. You can find plenty of books to tell you how radio sets operate and what to do, but very few except some for advanced students tell you how to experiment for yourself. Not to waste time in your own 4experiments, however, you will need to be quite familiar with the ideas of the other letters.

What is a radio set? Copper wires, tinfoil, glass plates, sheets of mica, metal, and wood. Where does it get its ability to work–that is, where does the "energy" come from which runs the set? From batteries or from dynamos. That much you know already, but what is the real reason that we can use copper wires, metal plates, audions, crystals, and batteries to send messages and to receive them?

The reason is that all these things are made of little specks, too tiny ever to see, which we might call specks of electricity. There are only two kinds of specks and we had better give them their right names at once to save time. One kind of speck is called "electron" and the other kind "proton." How do they differ? They probably differ in size but we don't yet know so very much about their sizes. They differ in laziness a great deal. One is about 1845 times as lazy as the other. That is, it has eighteen hundred and forty-five times as much inertia as the other. It is harder to get it started but it is just as much harder to get it to stop after it is once started or to change its direction and go a different direction. The proton has the larger inertia. It is the electron which is the easier to start or stop.

How else do they differ? They differ in their actions. Protons don't like to associate with other protons but take quite keenly to electrons. And electrons–they go with protons but they won't associate 5with each other. An electron always likes to be close to a proton. Two is company when one is an electron and the other a proton but three is a crowd always.

It doesn't make any difference to a proton what electron it is keeping company with provided only it is an electron and not another proton. All electrons are alike as far as we can tell and so are all protons. That means that all the stuff, or matter, of our world is made up of two kinds of building blocks, and all the blocks of each kind are just alike. Of course you mustn't think of these blocks as like bricks, for we don't know their shapes.

Then there is another reason why you must not think of them as bricks and that is because when you build a house out of bricks each brick must rest on another. Between an electron and any other electron or between two protons or between an electron and a proton there is usually a relatively enormous distance. There is enough space so that lots of other electrons or protons could be fitted in between if only they were willing to get that close together.

Sometimes they do get very close together. I can tell you how if you will imagine four small boys playing tag. Suppose Tom and Dick don't like to play with each other and run away from each other if they can. Now suppose that Bill and Sam won't play with each other if they can help it but that either of them will play with Tom or Dick whenever there is a chance. Now suppose Tom and Bill see 6each other; they start running toward each other to get up some sort of a game. But Sam sees Tom at the same time, so he starts running to join him even though Bill is going to be there too. Meanwhile Dick sees Bill and Sam running along and since they are his natural playmates he follows them. In a minute they are all together, and playing a great game; although some of the boys don't like to play together.

Whenever there is a group of protons and electrons playing together we have what we call an "atom." There are about ninety different games which electrons and protons can play, that is ninety different kinds of atoms. These games differ in the number of electrons and protons who play and in the way they arrange themselves. Larger games can be formed if a number of atoms join together. Then there is a "molecule." Of molecules there are as many kinds as there are different substances in the world. It takes a lot of molecules together to form something big enough to see, for even the largest molecule, that of starch, is much too small to be seen by itself with the best possible microscope.

What sort of a molecule is formed will depend upon how many and what kinds of atoms group together to play the larger game. Whenever there is a big game it doesn't mean that the little atomic groups which enter into it are all changed around. They keep together like a troop of boy scouts in a grand picnic in which lots of troops are present. At any rate they keep together enough so that we 7can still call them a group, that is an atom, even though they do adapt their game somewhat so as to fit in with other groups–that is with other atoms.

What will the kind of atom depend upon? It will depend upon how many electrons and protons are grouped together in it to play their little game. How any atom behaves so far as associating with other groups or atoms will depend upon what sort of a game its own electrons and protons are playing.

Now the simplest kind of a game that can be played, and the one with the smallest number of electrons and protons, is that played by a single proton and a single electron. I don't know just how it is played but I should guess that they sort of chase each other around in circles. At any rate I do know that the atom called "hydrogen" is formed by just one proton and one electron. Suppose they were magnified until they were as large as the moon and the earth. Then they would be just about as far apart but the smaller one would be the proton.

That hydrogen atom is responsible for lots of interesting things for it is a great one to join with other atoms. We don't often find it by itself although we can make it change its partners and go from one molecule to another very easily. That is what happens every time you stain anything with acid. A hydrogen atom leaves a molecule of the acid and then it isn't acid any more. What remains isn't a happy group either for it has lost some of its playfellows. The hydrogen goes and joins with the stuff which gets stained. But it doesn't join with the 8whole molecule; it picks out part of it to associate with and that leaves the other part to take the place of the hydrogen in the original molecule of acid from which it came. Many of the actions which we call chemistry are merely the result of such changes of atoms from one molecule to another.

Not only does the hydrogen atom like to associate in a larger game with other kinds of atoms but it likes to do so with one of its own kind. When it does we have a molecule of hydrogen gas, the same gas as is used in balloons.

We haven't seemed to get very far yet toward radio but you can see how we shall when I tell you that next time I shall write of more complicated games such as are played in the atoms of copper which form the wires of radio sets and of how these wires can do what we call "carrying an electric current."

9 LETTER 2
WHY A COPPER WIRE WILL CONDUCT ELECTRICITY

You have learned that the simplest group which can be formed by protons and electrons is one proton and one electron chasing each other around in a fast game. This group is called an atom of hydrogen. A molecule of hydrogen is two of these groups together.

All the other possible kinds of groups are more complicated. The next simplest is that of the atom of helium. Helium is a gas of which small quantities are obtained from certain oil wells and there isn't very much of it to be obtained. It is an inert gas, as we call it, because it won't burn or combine with anything else. It doesn't care to enter into the larger games of molecular groups. It is satisfied to be as it is, so that it isn't much use in chemistry because you can't make anything else out of it. That's the reason why it is so highly recommended for filling balloons or airships, because it cannot burn or explode. It is not as light as hydrogen but it serves quite well for making balloons buoyant in air.

This helium atom is made up of four electrons and four protons. Right at the center there is a small closely crowded group which contains all the protons 10and two of the electrons. The other two electrons play around quite a little way from this inner group. It will make our explanations easier if we learn to call this inner group "the nucleus" of the atom. It is the center of the atom and the other two electrons play around about it just as the earth and Mars and the other planets play or revolve about the sun as a center. That is why we shall call these two electrons "planetary electrons."

There are about ninety different kinds of atoms and they all have names. Some of them are more familiar than hydrogen and helium. For example, there is the iron atom, the copper atom, the sulphur atom and so on. Some of these atoms you ought to know and so, before telling you more of how atoms are formed by protons and electrons, I am going to write down the names of some of the atoms which we have in the earth and rocks of our world, in the water of the oceans, and in the air above.

Start first with air. It is a mixture of several kinds of gases. Each gas is a different kind of atom. There is just a slight trace of hydrogen and a very small amount of helium and of some other gases which I won't bother you with learning. Most of the air, however, is nitrogen, about 78 percent in fact and almost all the rest is oxygen. About 20.8 percent is oxygen so that all the gases other than these two make up only about 1.2 percent of the atmosphere in which we live.

Pl. II.–Bird's-eye View of Radio Central
(Courtesy of Radio Corporation of America).

11The earth and rocks also contain a great deal of oxygen; about 47.3 percent of the atoms which form earth and rocks are oxygen atoms. About half of the rest of the atoms are of a kind called silicon. Sand is made up of atoms of silicon and oxygen and you know how much sand there is. About 27.7 percent of the earth and its rocks is silicon. The next most important kind of atom in the earth is aluminum and after that iron and then calcium. Here is the way they run in percentages: Aluminum 7.8 percent; iron 4.5 percent; calcium 3.5 percent; sodium 2.4 percent; potassium 2.4 percent; magnesium 2.2 percent. Besides these which are most important there is about 0.2 percent of hydrogen and the same amount of carbon. Then there is a little phosphorus, a little sulphur, a little fluorine, and small amounts of all of the rest of the different kinds of atoms.

Sea water is mostly oxygen and hydrogen, about 85.8 percent of oxygen and 10.7 percent of hydrogen. That is what you would expect for water is made up of molecules which in turn are formed by two atoms of hydrogen and one atom of oxygen. The oxygen atom is about sixteen times as heavy as the hydrogen atom. However, for every oxygen atom there are two hydrogen atoms so that for every pound of hydrogen in water there are about eight pounds of oxygen. That is why there is about eight times as high a percentage of oxygen in sea water as there is of hydrogen.

Most of sea water, therefore, is just water, that is, pure water. But it contains some other substances as well and the best known of these is salt. Salt is a 12substance the molecules of which contain atoms of sodium and of chlorine. That is why sea water is about 1.1 percent sodium and about 2.1 percent chlorine. There are some other kinds of atoms in sea water, as you would expect, for it gets all the substances which the waters of the earth dissolve and carry down to it but they are unimportant in amounts.

Now we know something about the names of the important kinds of atoms and can take up again the question of how they are formed by protons and electrons. No matter what kind of atom we are dealing with we always have a nucleus or center and some electrons playing around that nucleus like tiny planets. The only differences between one kind of atom and any other kind are differences in the nucleus and differences in the number and arrangement of the planetary electrons which are playing about the nucleus.

No matter what kind of atom we are considering there is always in it just as many electrons as protons. For example, the iron atom is formed by a nucleus and twenty-six electrons playing around it. The copper atom has twenty-nine electrons as tiny planets to its nucleus. What does that mean about its nucleus? That there are twenty-nine more protons in the nucleus than there are electrons. Silver has even more planetary electrons, for it has 47. Radium has 88 and the heaviest atom of all, that of uranium, has 92.

We might use numbers for the different kinds of 13atoms instead of names if we wanted to do so. We could describe any kind of atom by telling how many planetary electrons there were in it. For example, hydrogen would be number 1, helium number 2, lithium of which you perhaps never heard, would be number 3, and so on. Oxygen is 8, sodium is 11, chlorine is 17, iron 26, and copper 29. For each kind of atom there is a number. Let's call that number its atomic number.

Now let's see what the atomic number tells us. Take copper, for example, which is number 29. In each atom of copper there are 29 electrons playing around the nucleus. The nucleus itself is a little inner group of electrons and protons, but there are more protons than electrons in it; twenty-nine more in fact. In an atom there is always an extra proton in the nucleus for each planetary electron. That makes the total number of protons and electrons the same.

About the nucleus of a copper atom there are playing 29 electrons just as if the nucleus was a teacher responsible for 29 children who were out in the play yard. There is one very funny thing about it all, however, and that is that we must think of the scholars as if they were all just alike so that the teacher couldn't tell one from the other. Electrons are all alike, you remember. All the teacher or nucleus cares for is that there shall be just the right number playing around her. You could bring a boy in from some other play ground and the teacher couldn't tell that he was a stranger but she would 14know that something was the matter for there would be one too many in her group. She is responsible for just 29 scholars, and the nucleus of the copper atom is responsible for just 29 electrons. It doesn't make any difference where these electrons come from provided there are always just 29 playing around the nucleus. If there are more or less than 29 something peculiar will happen.

We shall see later what might happen, but first let's think of an enormous lot of atoms such as there would be in a copper wire. A small copper wire will have in it billions of copper atoms, each with its planetary electrons playing their invisible game about their own nucleus. There is quite a little distance in any atom between the nucleus and any of the electrons for which it is responsible. There is usually a greater distance still between one atomic group and any other.

On the whole the electrons hold pretty close to their own circles about their own nuclei. There is always some tendency to run away and play in some other group. With 29 electrons it's no wonder if sometimes one goes wandering off and finally gets into the game about some other nucleus. Of course, an electron from some other atom may come wandering along and take the place just left vacant, so that nucleus is satisfied.

We don't know all we might about how the electrons wander around from atom to atom inside a copper wire but we do know that there are always a lot of them moving about in the spaces between 15the atoms. Some of them are going one way and some another.

It's these wandering electrons which are affected when a battery is connected to a copper wire. Every single electron which is away from its home group, and wandering around, is sent scampering along toward the end of the wire which is connected to the positive plate or terminal of the battery and away from the negative plate. That's what the battery does to them for being away from home; it drives them along the wire. There's a regular stream or procession of them from the negative end of the wire toward the positive. When we have a stream of electrons like this we say we have a current of electricity.

We'll need to learn more later about a current of electricity but one of the first things we ought to know is how a battery is made and why it affects these wandering electrons in the copper wire. That's what I shall tell you in my next letter. [1]

[1]

The reader who wishes the shortest path to the construction and operation of a radio set should omit the next two letters.

16 LETTER 3
HOW A BATTERY WORKS

(This letter may be omitted on the first reading.)

MY DEAR BOY:

When I was a boy we used to make our own batteries for our experiments. That was before storage batteries became as widely used as they are to-day when everybody has one in the starting system of his automobile. That was also before the day of the small dry battery such as we use in pocket flash lights. The batteries which we made were like those which they used on telegraph systems, and were sometimes called "gravity" batteries. Of course, we tried several kinds and I believe I got quite a little acid around the house at one time or another. I'll tell you about only one kind but I shall use the words "electron," "proton," "nucleus," "atom," and "molecule," about some of which nothing was known when I was a boy.

We used a straight-sided glass jar which would hold about a gallon. On the bottom we set a star shaped arrangement made of sheets of copper with a long wire soldered to it so as to reach up out of the jar. Then we poured in a solution of copper sulphate until the jar was about half full. This solution was made by dissolving in water crystals of "blue vitriol" which we bought at the drug store.

17Blue vitriol, or copper sulphate as the chemists would call it, is a substance which forms glassy blue crystals. Its molecules are formed of copper atoms, sulphur atoms, and oxygen atoms. In each molecule of it there is one atom of copper, one of sulphur and four of oxygen.

When it dissolves in water the molecules of the blue vitriol go wandering out into the spaces between the water molecules. But that isn't all that happens or the most important thing for one who is interested in making a battery.

Each molecule is formed by six atoms, that is by six little groups of electrons playing about six little nuclei. About each nucleus there is going on a game but some of the electrons are playing in the game about their own nucleus and at the same time taking some part in the game which is going on around one of the other nuclei. That's why the groups or atoms stay together as a molecule. When the molecules wander out into the spaces between the water molecules something happens to this complicated game.

It will be easiest to see what sort of thing happens if we talk about a molecule of ordinary table salt, for that has only two atoms in it. One atom is sodium and one is chlorine. The sodium molecule has eleven electrons playing around its nucleus. Fairly close to the nucleus there are two electrons. Then farther away there are eight more and these are having a perfect game. Then still farther away from the nucleus there is a single lonely electron.

The atom of chlorine has seventeen electrons which 18play about its nucleus. Close to the nucleus there are two. A little farther away there are eight just as there are in the sodium atom. Then still farther away there are seven.

I am going to draw a picture (Fig. 1) to show what I mean, but you must remember that these electrons are not all in the same plane as if they lay on a sheet of paper, but are scattered all around just as they would be if they were specks on a ball.

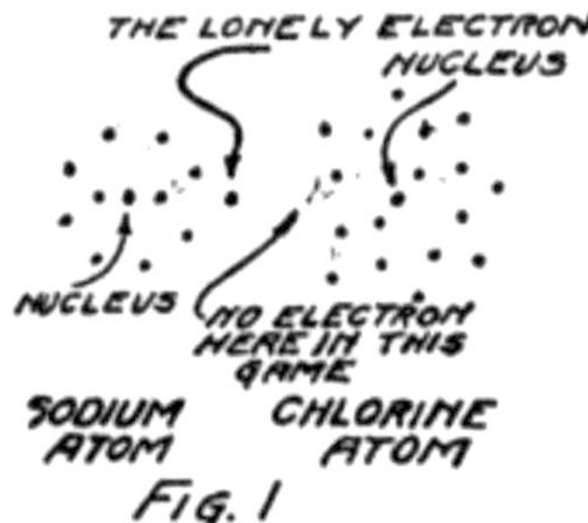

FIG. 1

You see that the sodium atom has one lonely electron which hasn't any play fellows and that the chlorine atom has seven in its outside circle. It appears that eight would make a much better game. Suppose that extra electron in the sodium atom goes over and plays with those in the chlorine atom so as to make eight in the outside group as I have shown Fig. 2. That will be all right as long as it doesn't get out of sight of its own nucleus because you remember that the sodium nucleus is responsible for eleven electrons. The lonely electron of the sodium atom needn't be lonely any more if it can persuade its nucleus to stay so close to the chlorine atom that it can play in the outer circle of the chlorine atom.

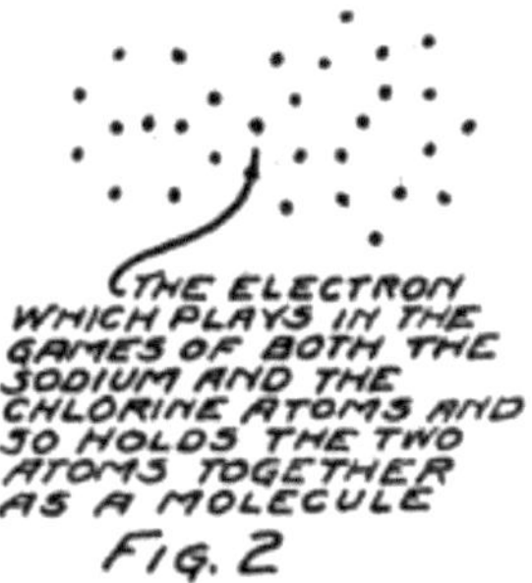

FIG. 2

19The outer circle of the chlorine atom will then have a better game, for it will have just the eight that makes a perfect game. This can happen if the chlorine atom will stay close enough to the sodium atom so that the outermost electron of the sodium atom can play in the chlorine circle. You see everything will be satisfactory if an electron can be shared by the two atoms. That can happen only if the two atoms stay together; that is, if they form a molecule. That's why there are molecules and that's what I meant when I spoke of the molecule as a big game played by the electrons of two or more atoms.

This molecule which is formed by a sodium atom and a chlorine atom is called a molecule of sodium chloride by chemists and a molecule of salt by most every one who eats it. Something strange happens when it dissolves. It wanders around between the water molecules and for some reason or other–we don't know exactly why–it decides to split up again into sodium and chlorine but it can't quite do it. The electron which joined the game about the chlorine nucleus won't leave it. The result is that the nucleus of the sodium atom gets away but it leaves this one electron behind.

What gets away isn't a sodium atom for it has one too few electrons; and what remains behind isn't a chlorine atom for it has one too many electrons. We call these new groups "ions" from a Greek word which means "to go" for they do go, wandering off into the spaces between the water 20molecules. Fig. 3 gives you an idea of what happens.

You remember that in an atom there are always just as many protons as electrons. In this sodium ion which is formed when the nucleus of the sodium atom breaks away but leaves behind one planetary electron, there is then one more proton than there are electrons. Because it has an extra proton, which hasn't any electron to associate with, we call it a plus ion or a "positive ion." Similarly we call the chlorine ion, which has one less proton than it has electrons, a minus or "negative ion."

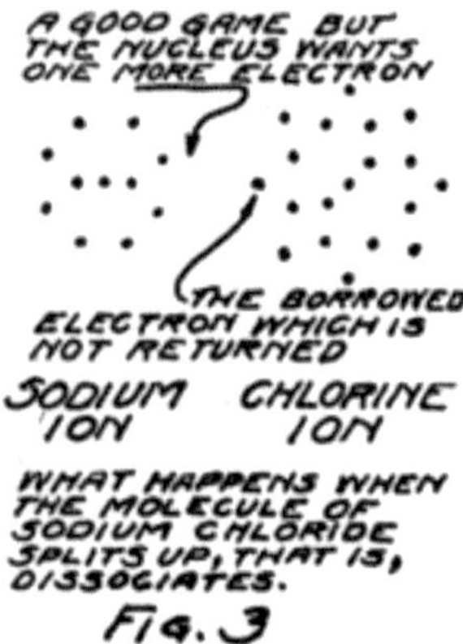

Now, despite the fact that these ions broke away from each other they aren't really satisfied. Any time that the sodium ion can find an electron to take the place of the one it lost it will welcome it. That is, the sodium ion will want to go toward places where there are extra electrons. In the same way the chlorine ion will go toward places where electrons are wanted as if it could satisfy its guilty conscience by giving up the electron which it stole from the sodium atom, or at least by giving away some other electron, for they are all alike anyway.

Sometimes a positive sodium ion and a negative chlorine ion meet in their wanderings in the solution and both get satisfied by forming a molecule 21again. Even so they don't stay together long before they split apart and start wandering again. That's what goes on over and over again, millions of times, when you dissolve a little salt in a glass of water.

Now we can see what happens when copper sulphate dissolves. The copper atom has twenty-nine electrons about its nucleus and all except two of these are nicely grouped for playing their games about the nucleus. Two of the electrons are rather out of the game, and are unsatisfied. They play with the electrons of the part of the molecule which is called "sulphate," that is, the part formed by the sulphur atom and the four oxygen atoms. These five atoms of the sulphate part stay together very well and so we treat them as a group.

The sulphate group and the copper atom stay together as long as they are not in solution but when they are, they act very much like the sodium and chlorine which I just described. The molecule splits up into two ions, one positive and one negative. The positive ion is the copper part except that two of the electrons which really belong to a copper atom got left behind because the sulphate part wouldn't give them up. The rest of the molecule is the negative ion.

The copper ion is a copper atom which has lost two electrons. The sulphate ion is a combination of one sulphur atom, four oxygen atoms and two electrons which it stole from the copper atom. Just as the sodium ion is unsatisfied because in it there is one more proton than there are electrons, so the copper ion is unsatisfied. As a matter of fact it is twice 22as badly unsatisfied. It has two more protons than it has electrons. We say it has twice the "electrical charge" of the sodium ion.

Just like a sodium ion the copper ion will tend to go toward any place where there are extra electrons which it can get to satisfy its own needs. In much the same way the sulphate ion will go toward places where it can give up its two extra electrons. Sometimes, of course, as ions of these two kinds wander about between the water molecules, they meet and satisfy each other by forming a molecule of copper sulphate. But if they do they will split apart later on; that is, they will "dissociate" as we should say.

Now let's go on with the kind of batteries I used to make as a boy. You can see that in the solution of copper sulphate at the bottom of the jar there was always present a lot of positive copper ions and of negative sulphate ions.

On top of this solution of copper sulphate I poured very carefully a weak solution of sulphuric acid. As I told you, an acid always has hydrogen in its molecules. Sulphuric acid has molecules formed by two hydrogen atoms and one of the groups which we decided to call sulphate. A better name for this acid would be hydrogen sulphate for that would imply that its molecule is the same as one of copper sulphate, except that the place of the copper is taken by two atoms of hydrogen. It takes two atoms of hydrogen because the copper atom has two lonely electrons while a hydrogen atom only has one. It takes two electrons to fill up the game which the 23electrons of the sulphate group are playing. If it can get these from a single atom, all right; but if it has to get one from each of two atoms, it will do it that way.

I remember when I mixed the sulphuric acid with water that I learned to pour the acid into the water and not the other way around. Spatterings of sulphuric acid are not good for hands or clothes. With this solution I filled the jar almost to the top and then hung over the edge a sort of a crow's foot shape of cast zinc. The zinc reached down into the sulphuric acid solution. There was a binding post on it to which a wire could be connected. This wire and the one which came from the plate of copper at the bottom were the two terminals of the battery. We called the wire from the copper "positive" and the one from the zinc "negative."

Now we shall see why and how the battery worked. The molecules of sulphuric acid dissociate in solution just as do those of copper sulphate. When sulphuric acid molecules split, the sulphate part goes away with two electrons which don't belong to it and each of the hydrogen atoms goes away by itself but without its electron. We call each a "hydrogen ion" but you can see that each is a single proton.

In the two solutions are pieces of zinc and copper. Zinc is like all the rest of the metals in one way. Atoms of metals always have lonely electrons for which there doesn't seem to be room in the game which is going on around their nuclei. Copper as we saw has two lonely electrons in each atom. Zinc 24also has two. Some metals have one and some two and some even more lonely electrons in each atom.

What happens then is this. The sulphate ions wandering around in the weak solution of sulphuric acid come along beside the zinc plate and beckon to its atoms. The sulphate ions had a great deal rather play the game called "zinc sulphate" than the game called "hydrogen sulphate." So the zinc atoms leave their places to join with the sulphate ions. But wait a minute! The sulphate ions have two extra electrons which they kept from the hydrogen atoms. They don't need the two lonely electrons which each zinc atom could bring and so the zinc atom leaves behind it these unnecessary electrons.

Every time a zinc atom leaves the plate it fails to take all its electrons with it. What leaves the zinc plate, therefore, to go into solution is really not a zinc atom but is a zinc ion; that is, it is the nucleus of a zinc atom and all except two of the planetary electrons.

Every time a zinc ion leaves the plate there are left behind two electrons. The plate doesn't want them for all the rest of its atoms have just the same number of protons as of electrons. Where are they to go? We shall see in a minute.

Sometimes the zinc ions which have got into solution meet with sulphate ions and form zinc sulphate molecules. But if they do these molecules split up sooner or later into ions again. In the upper part of the liquid in the jar, therefore, there are sulphate 25ions which are negative and two kinds of positive ions, namely, the hydrogen ions and the zinc ions.

Before the zinc ions began to crowd in there were just enough hydrogen ions to go with the sulphate ions. As it is, the entrance of the zinc ions has increased the number of positive ions and now there are too many. Some of the positive ions, therefore, and particularly the hydrogen ions, because the sulphate prefers to associate with the zinc ions, can't find enough playfellows and so go down in the jar.

Down in the bottom of the jar the hydrogen ions find more sulphate ions to play with, but that leaves the copper ions which used to play with these sulphate ions without any playmates. So the copper ions go still further down and join with the copper atoms of the copper plate. They haven't much right to do so, for you remember that they haven't their proper number of electrons. Each copper ion lacks two electrons of being a copper atom. Nevertheless they join the copper plate. The result is a plate of copper which has too few electrons. It needs two electrons for every copper ion which joins it.

How about the zinc plate? You remember that it has two electrons more than it needs for every zinc ion which has left it. If only the extra electrons on the negative zinc plate could get around to the positive copper plate. They can if we connect a wire from one plate to the other. Then the electrons from the zinc stream into the spaces between the atoms of the wire and push ahead of them the electrons 26which are wandering around in these spaces. At the other end an equal number of electrons leave the wire to satisfy the positive copper plate. So we have a stream of electrons in the wire, that is, a current of electricity and our battery is working.

That's the sort of a battery I used to play with. If you understand it you can get the general idea of all batteries. Let me express it in general terms.

At the negative plate of a battery ions go into solution and electrons are left behind. At the other end of the battery positive ions are crowded out of solution and join the plate where they cause a scarcity of electrons; that is, make the plate positive. If a wire is connected between the two plates, electrons will stream through it from the negative plate to the positive; and this stream is a current of electricity.

EVEREADY

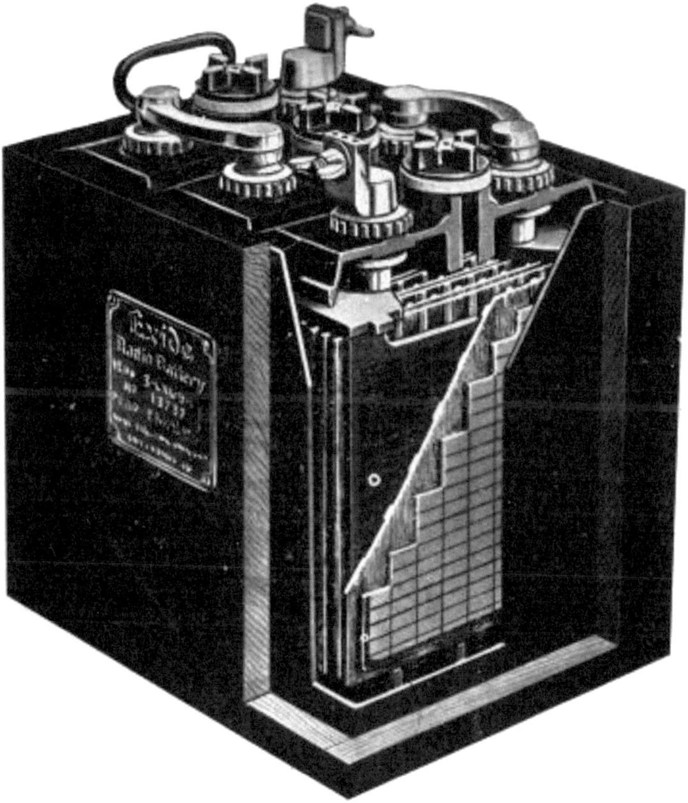

Pl. III.–Dry Battery for Use in Audion Circuits (Courtesy of National Carbon Co., Inc.)

Storage Battery (Courtesy of the Electric Storage Battery Co.).

32

27 LETTER 4
THE BATTERIES IN YOUR RADIO SET

(This letter may be omitted on the first reading.)

MY DEAR YOUNG MAN:

You will need several batteries when you come to set up your radio receiver but you won't use such clumsy affairs as the gravity cell which I described in my last letter. Some of your batteries will be dry batteries of the size used in pocket flash lights.

These are not really dry, for between the plates they are filled with a moist paste which is then sealed in with wax to keep it from drying out or from spilling. Instead of zinc and copper these batteries use zinc and carbon. No glass jar is needed, for the zinc is formed into a jar shape. In this is placed the paste and in the center of the paste a rod or bar of carbon. The paste doesn't contain sulphuric acid, but instead has in it a stuff called sal ammoniac; that is, ammonium chloride.

The battery, however, acts very much like the one I described in my last letter. Ions of zinc leave the zinc and wander into the moist paste. These ions are positive, just as in the case of the gravity battery. The result is that the electrons which used to associate with a zinc ion to form a zinc atom are left in the zinc plate. That makes the zinc negative 28for it has more electrons than protons. The zinc ions take the place of the positive ions which are already in the paste. The positive ions which originally belonged with the paste, therefore, move along to the carbon rod and there get some electrons. Taking electrons away from the carbon leaves it with too many protons; that is, leaves it positive. In the little flash light batteries, therefore, you will always find that the round carbon rod, which sticks out of the center, is positive and the zinc casing is negative.

The trouble with the battery like the one I used to make is that the zinc plate wastes away. Every time a zinc ion leaves it that means that the greater part of an atom is gone. Then when the two electrons which were left behind get a chance to start along a copper wire toward the positive plate of the battery there goes the rest of the atom. After a while there is no more zinc plate. It is easy to see what has happened. All the zinc has gone into solution or been "eaten away" as most people say. Dry batteries, however, don't stop working because the zinc gets used up, but because the active stuff in the paste, the ammonium chloride, is changed into something else.

There's another kind of battery which you will need to use with your radio set; that is the storage battery. Storage batteries can be used over and over again if they are charged between times and will last for a long time if properly cared for. Then too, they can give a large current, that is, a big swift-moving stream of electrons. You will need 29that when you wish to heat the filament of the audion in your receiving set.

The English call our storage batteries by the name "accumulators." I don't like that name at all, but I don't like our name for them nearly as well as I do the name "reversible batteries." Nobody uses this last name because it's too late to change. Nevertheless a storage battery is reversible, for it will work either way at an instant's notice.

A storage battery is something like a boy's wagon on a hill side. It will run down hill but it can be pushed up again for another descent. You can use it to send a stream of electrons through a wire from its negative plate to its positive plate. Then if you connect these plates to some other battery or to a generator, (that is, a dynamo) you can make a stream of electrons go in the other direction. When you have done so long enough the battery is charged again and ready to discharge.

I am not going to tell you very much about the storage battery but you ought to know a little about it if you are to own and run one with your radio set. When it is all charged and ready to work, the negative plate is a lot of soft spongy lead held in place by a frame of harder lead. The positive plate is a lead frame with small squares which are filled with lead peroxide, as it is called. This is a substance with molecules formed of one lead atom and two oxygen atoms. Why the chemists call it lead peroxide instead of just lead oxide I'll tell you some other time, but not in these letters.

30Between the two plates is a wood separator to keep pieces of lead from falling down between and touching both plates. You know what would happen if a piece of metal touched both plates. There would be a short circuit, that is, a sort of a short cut across lots by which some of the electrons from the negative plate could get to the positive plate without going along the wires which we want them to travel. That's why there are separators.

The two plates are in a jar of sulphuric acid solution. The sulphuric acid has molecules which split up in solution, as you remember, into hydrogen ions and the ions which we called "sulphate." In my gravity battery the sulphate ions used to coax the zinc ions away into the solution. In the storage battery on the other hand the sulphate ions can get to most of the lead atoms because the lead is so spongy. When they do, they form lead sulphate right where the lead atoms are. They don't really need whole lead atoms, because they have two more electrons than they deserve, so there are two extra electrons for every molecule of lead sulphate which is formed. That's why the spongy lead plate is negative.

The lead sulphate won't dissolve, so it stays there on the plate as a whitish coating. Now see what that means. What are the hydrogen ions going to do? As long as there was sulphuric acid in the water there was plenty of sulphate ions for them to associate with as often as they met; and they would meet pretty often. But if the sulphate ions get tied up 31with the lead of the plate there will be too many hydrogen ions left in the solution. Now what are the hydrogen ions to do? They are going to get as far away from each other as they can, for they are nothing but protons; and protons don't like to associate. They only stayed around in the first place because there was always plenty of sulphate ions with whom they liked to play.

When the hydrogen ions try to get away from each other they go to the other plate of the battery, and there they will get some electrons, if they have to steal in their turn.

I won't try to tell you all that happens at the other plate. The hydrogen ions get the electrons which they need, but they get something more. They get some of the oxygen away from the plate and so form molecules of water. You remember that water molecules are made of two atoms of hydrogen and one of oxygen. Meanwhile, the lead atoms, which have lost their oxygen companions, combine with some of the sulphate ions which are in that neighborhood.

During the mix-up electrons are carried away from the plate and that leaves it positive.

The result of all this is a little lead sulphate on each plate, a negative plate where the spongy lead was, and a positive plate where the lead peroxide was.

Notice very carefully that I said "a little lead sulphate on each plate." The sort of thing I have been describing doesn't go on very long. If it did the 32battery would run down inside itself and then when we came to start our automobile we would have to get out and crank.

How long does it go on? Answer another question first. So far we haven't connected any wire between the two plates of the battery, and so none of the electrons on the negative plate have any way of getting around to the positive plate where electrons are badly needed. Every time a negative sulphate ion combines with the spongy lead of the negative plate there are two more electrons added to that plate. You know how well electrons like each other. Do they let the sulphate ions keep giving that plate more electrons? There is the other question; and the answer is that they do not. Every electron that is added to that plate makes it just so much harder for another sulphate ion to get near enough to do business at all. That's why after a few extra electrons have accumulated on the spongy lead plate the actions which I was describing come to a stop.

Do they ever begin again? They do just as soon as there is any reduction in the number of electrons which are hopping around in the negative plate trying to keep out of each other's way. When we connect a wire between the plates we let some of these extra electrons of the negative plate pass along to the positive plate where they will be welcome. And the moment a couple of them start off on that errand along comes another sulphate ion in the solution and lands two more electrons on the plate. That's how the battery keeps on discharging.

33We mustn't let it get too much discharged for the lead sulphate is not soluble, as I just told you, and it will coat up that plate until there isn't much chance of getting the process to reverse. That's why we are so careful not to let the discharge process go on too long before we reverse it and charge. That's why, when the car battery has been used pretty hard to start the car, I like to run quite a while to let the generator charge the battery again. When the battery charges, the process reverses and we get spongy lead on the negative plate and lead peroxide on the positive plate.

You've learned enough for one day. Write me your questions and I'll answer and then go on in my next letter to tell how the audion works. You know about conduction of electricity in wires; that is, about the electron stream, and about batteries which can cause the stream. Now you are ready for the most wonderful little device known to science: the audion.

34 LETTER 5
GETTING ELECTRONS FROM A HEATED WIRE

Dear Son:

I was pleased to get your letter and its questions. Yes, a proton is a speck of electricity of the kind we call positive and an electron is of the kind we call negative. You might remember this simple law; "Like kinds of electricity repel, and unlike attract."

The word ion [2] is used to describe any atom, or part of a molecule which can travel by itself and has more or less than its proper number of electrons. By proper number of electrons I mean proper for the number of protons which it has. If an ion has more electrons than protons it is negative; if the inequality is the other way around it is positive. An atom or molecule has neither more nor less protons than electrons. It is neutral or "uncharged," as we say.

No, not every substance which will dissolve will dissociate or split up into positive and negative ions. The salt which you eat will, but the sugar will not. If you want a name for those substances which will dissociate in solution, call them "electrolytes." To make a battery we must always use an electrolyte.

Yes, it is hard to think of a smooth piece of metal or a wire as full of holes. Even in the densest solids like lead the atoms are quite far apart and there are 35large spaces between the nuclei and the planetary electrons of each atom.

I hope this clears up the questions in your mind for I want to get along to the vacuum tube. By a vacuum we mean a space which has very few atoms or molecules in it, just as few as we can possibly get, with the best methods of pumping and exhausting. For the present let's suppose that we can get all the gas molecules, that is, all the air, out of a little glass bulb.

The audion is a glass bulb like an electric light bulb which has in it a thread, or filament, of metal. The ends of this filament extend out through the glass so that we may connect a battery to them and pass a current of electricity through the wire. If we do so the wire gets hot.

What do we mean when we say "the wire gets hot?" We mean that it feels hot. It heats the glass bulb and we can feel it. But what do we mean in words of electrons and atoms? To answer this we must start back a little way.

In every bit of matter in our world the atoms and molecules are in very rapid motion. In gases they can move anywhere; and do. That's why odors travel so fast. In liquids most of the molecules or atoms have to do their moving without getting out of the dish or above the surface. Not all of them stay in, however, for some are always getting away from the liquid and going out into the air above. That is why a dish of water will dry up so quickly. The faster the molecules are going the better chance 36they have of jumping clear away from the water like fish jumping in the lake at sundown. Heating the liquid makes its molecules move faster and so more of them are able to jump clear of the rest of the liquid. That's why when we come in wet we hang our clothes where they will get warm. The water in them evaporates more quickly when it is heated because all we mean by "heating" is speeding up the molecules.

In a solid body the molecules can't get very far away from where they start but they keep moving back and forth and around and around. The hotter the body is, the faster are its molecules moving. Generally they move a little farther when the body is hot than when it is cold. That means they must have a little more room and that is why a body is larger when hot than when cold. It expands with heating because its molecules are moving more rapidly and slightly farther.

When a wire is heated its molecules and atoms are hurried up and they dash back and forth faster than before. Now you know that a wire, like the filament of a lamp, gets hot when the "electricity is turned on," that is, when there is a stream of electrons passing through it. Why does it get hot? Because when the electrons stream through it they bump and jostle their way along like rude boys on a crowded sidewalk. The atoms have to step a bit more lively to keep out of the way. These more rapid motions of the atoms we recognize by the wire growing hotter.

37That is why an electric current heats a wire through which it is flowing. Now what happens to the electrons, the rude boys who are dodging their way along the sidewalk? Some of them are going so fast and so carelessly that they will have to dodge out into the gutter and off the sidewalk entirely. The more boys that are rushing along and the faster they are going the more of them will be turned aside and plunge off the sidewalks.

The greater and faster the stream of electrons, that is the more current which is flowing through the wire, the more electrons will be "emitted," that is, thrown out of the wire. If you could watch them you would see them shooting out of the wire, here, there, and all along its length, and going in every direction. The number which shoot out each second isn't very large until they have stirred things up so that the wire is just about red hot.

What becomes of them? Sometimes they don't get very far away from the wire and so come back inside again. They scoot off the sidewalk and on again just as boys do in dodging their way along. Some of them start away as if they were going for good.

If the wire is in a vacuum tube, as it is in the case of the audion, they can't get very far away. Of course there is lots of room; but they are going so fast that they need more room just as older boys who run fast need a larger play ground than do the little tots. By and by there gets to be so many of them outside that they have to dodge each other and some of them are always dodging back into the 38wire while new electrons are shooting out from it.

When there are just as many electrons dodging back into the wire each second as are being emitted from it the vacuum in the tube has all the electrons it can hold. We might say it is "saturated" with electrons, which means, in slang, "full up." If any more electrons are to get out of the filament just as many others which are already outside have to go back inside. Or else they have got to be taken away somewhere else.

What I have just told you about electrons getting away from a heated wire is very much like what happens when a liquid is heated. The molecules of the liquid get away from the surface. If we cover a dish of liquid which is being heated the liquid molecules can't get far away and very soon the space between the surface of the liquid and the cover gets saturated with them. Then every time another molecule escapes from the surface of the liquid there must be some molecule which goes back into the liquid. There is then just as much condensation back into liquid as there is evaporation from it. That's why in cooking they put covers over the vessels when they don't want the liquid all to "boil away."

Sometimes we speak of the vacuum tube in the same words we would use in describing evaporation of a liquid. The molecules of the liquid which have escaped form what is called a "vapor" of the liquid. As you know there is usually considerable water vapor in the air. We say then that electrons are 39"boiled out" of the filament and that there is a "vapor of electrons" in the tube.

That is enough for this letter. Next time I shall tell you how use is made of these electrons which have been boiled out and are free in the space around the filament.

[2]

If the reader has omitted Letters 3 and 4 he should omit this paragraph and the next.

40 LETTER 6
THE AUDION

Dear Son:

In my last letter I told how electrons are boiled out of a heated filament. The hotter the filament the more electrons are emitted each second. If the temperature is kept steady, or constant as we say, then there are emitted each second just the same number of electrons. When the filament is enclosed in a vessel or glass bulb these electrons which get free from it cannot go very far away. Some of them, therefore, have to come back to the filament and the number which returns each second is just equal to the number which is leaving. You realize that this is what is happening inside an ordinary electric light bulb when its filament is being heated.

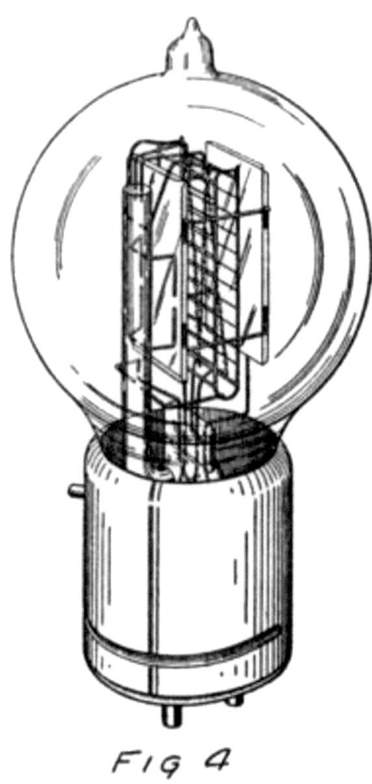

41An ordinary electric light bulb, however, is not an audion although it is like one in the emission of electrons from its filament. That reminds me that last night as I was waiting for a train I picked up one of the Radio Supplements which so many newspapers are now running. There was a column of enquiries. One letter told how its writer had tried to use an ordinary electric light bulb to receive radio signals.

He had plenty of electrons in it but no way to control them and make their motions useful. In an audion besides the filament there are two other things. One is a little sheet or plate of metal with a connecting wire leading out through the glass walls and the other is a little wire screen shaped like a gridiron and so called a "grid." It also has a connecting wire leading through the glass. Fig. 4 shows an audion. It will be most convenient, however, to represent an

audion as in Fig. 5. There you see the filament, *F*, with its two terminals brought out from the tube, the plate, *P*, and between these
the grid, *G*.

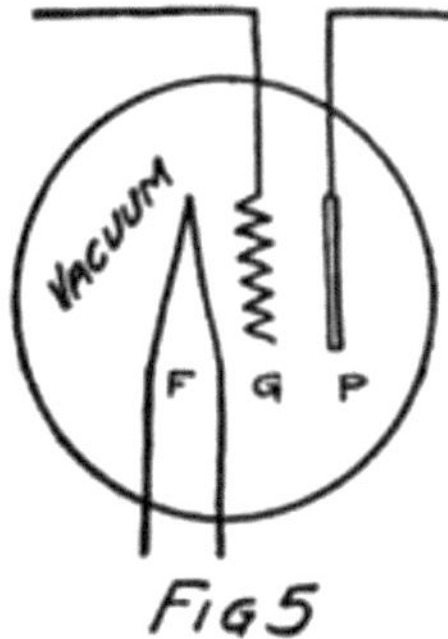

Fig 5

These three parts of the tube are sometimes called "elements."
Usually, however, they are called "electrodes" and that is why the
audion is spoken of as the "three-electrode vacuum tube." An elec-
trode is what we call any piece of metal or wire which is so placed
as to let us get at electrons (or 42ions) to control their motions. Let
us see how it does so.

To start with, we shall forget the grid and think of a tube with only a filament and a plate in it–a two-electrode tube. We shall represent it as in Fig. 6 and show the battery which heats the filament by
some lines as at *A*. In this way of representing a battery each cell is
represented by a short heavy line and a longer lighter line. The
heavy line stands for the negative plate and the longer line for the
positive plate. We shall call the battery which heats the filament the
"filament battery" or sometimes the "A-battery." As you see, it is
formed by several battery cells connected in series.

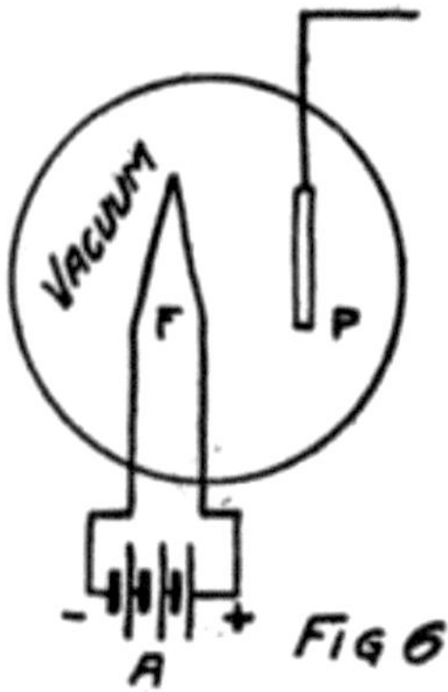

Sometime later I may tell you how to connect battery cells together and why. For the present all you need to remember is that two batteries are in series if the positive plate of one is connected to the negative plate of the other. If the batteries are alike they will pull an electron just twice as hard as either could alone.

43To heat the filament of an audion, such as you will probably use in your set, will require three storage-battery cells, like the one I described in my fourth letter, all connected in series. We generally use storage batteries of about the same size as those in the automobile. If you will look at the automobile battery you will see that it is made of three cells connected in series. That battery would do very well for the filament circuit.

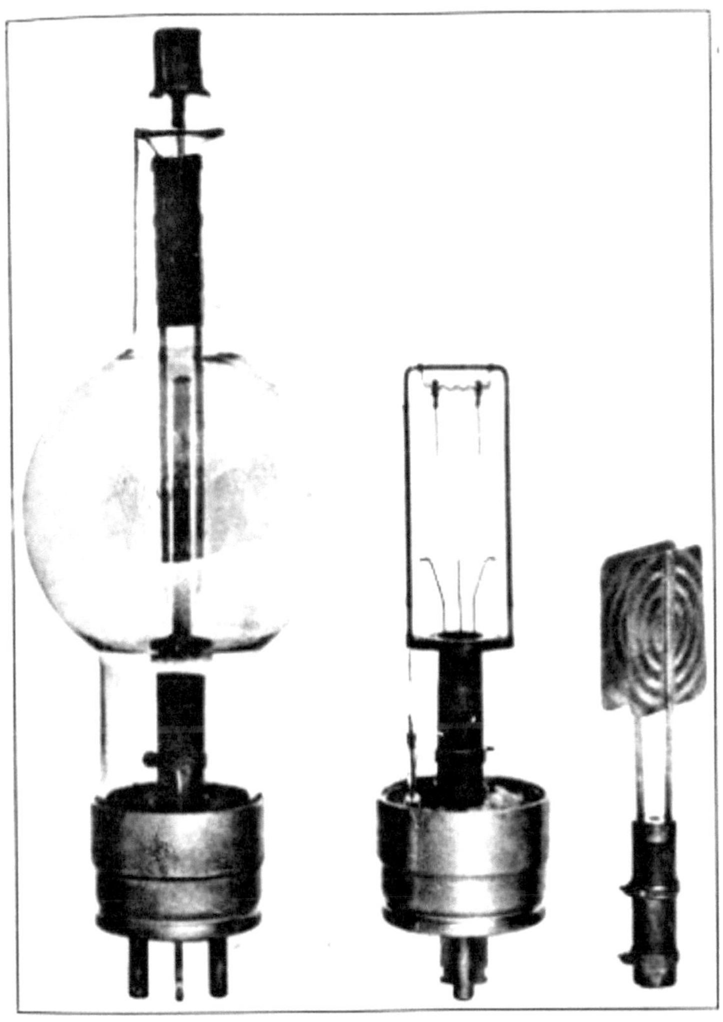

Pl. IV.–Radiotron (Courtesy of Radio Corporation of America).

44By the way, do you know what a "circuit" is? The word comes from the same Latin word as our word "circus." The Romans were very fond of chariot racing at their circuses and built race tracks around which the chariots could go. A circuit, therefore, is a path or track around which something can race; and an electrical circuit is a path around which electrons can race. The filament, the A-battery and the connecting wires of Fig. 6 form a circuit.

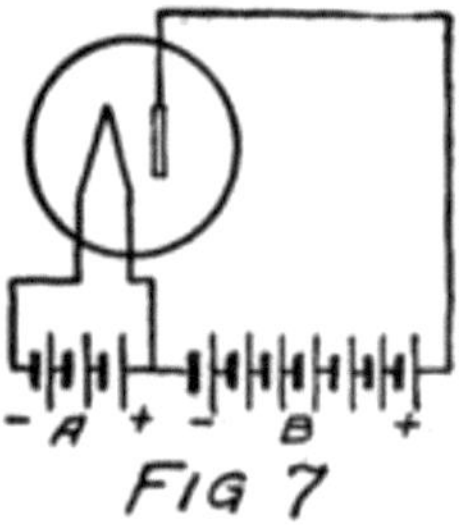

Let us imagine another battery formed by several cells in series which we shall connect to the tube as in Fig. 7. All the positive and negative terminals of these batteries are connected in pairs, the positive of one cell to the negative of the next, except for one positive and one negative. The remaining positive terminal is the positive terminal of the battery which we are making by this series connection. We then connect this positive terminal to the plate and the negative terminal to the filament as shown in the figure. This new battery we shall call the "plate battery" or the "B-battery."

Now what's going to happen? The B-battery will want to take in electrons at its positive terminal and to send them out at its negative terminal. The positive is connected to the plate in the vacuum tube 45of the figure and so draws some of the electrons of the plate away from it. Where do these electrons come from? They used to belong to the atoms of the plate but they were out playing in the space between the atoms, so that they came right along when the battery called them. That leaves the plate with less than its proper number of electrons; that is, leaves it positive. So the plate immediately draws to itself some of the electrons which are dodging about in the vacuum around it.

Do you remember what was happening in the tube? The filament was steadily going on emitting electrons although there were already in the tube so many electrons that just as many crowded back into the filament each second as the filament sent out. The filament was neither gaining nor losing electrons, although it was busy sending them out and welcoming them home again.

When the B-battery gets to work all this is changed. The B-battery attracts electrons to the plate and so reduces the crowd in the tube. Then there are not as many electrons crowding back into the filament as there were before and so the filament loses more than it gets back.

Suppose that, before the B-battery was connected to the plate, each tiny length of the filament was emitting 1000 electrons each second but was getting 1000 back each second. There was no net change. Now, suppose that the B-battery takes away 100 of these each second. Then only 900 get back to the filament and there is a net loss from the filament 46of 100. Each second this tiny length of filament sends into the vacuum 100 electrons which are taken out at the plate. From each little bit of filament there is a stream of electrons to the plate. Millions of electrons, therefore, stream across from filament to plate. That is, there is a current of electricity between filament and plate and this current continues to flow as long as the A-battery and the B-battery do their work.

The negative terminal of the B-battery is connected to the filament. Every time this battery pulls an electron from the plate its negative terminal shoves one out to the filament. You know from my third and fourth letters that electrons are carried through a battery from its positive to its negative terminal. You see, then, that there is the same stream of electrons through the B-battery as there is through the vacuum between filament and plate. This same stream passes also through the wires which connect the battery to the tube. The path followed by the stream of electrons includes the wires, the vacuum and the battery in series. We call this path the "plate circuit."

We can connect a telephone receiver, or a current-measuring instrument, or any thing we wish which will pass a stream of electrons, so as to let this same stream of electrons pass through it also. All we have to do is to connect the instrument in series with the other parts of the plate circuit. I'll show you how in a minute, but just now I want you to understand that we have a stream of electrons, 47for I want to tell you how it may be controlled.

Suppose we use another battery and connect it between the grid and the filament so as to make the grid positive. That would mean connecting the positive terminal of the battery to the grid and the negative to the filament as shown by the C-battery of Fig. 8. This figure also shows a current-measuring instrument in the plate circuit.

What effect is this C-battery, or grid-battery, going to have on the current in the *plate circuit*? Making the grid positive makes it want electrons. It will therefore act just as we saw that the plate did and pull electrons across the vacuum towards itself.

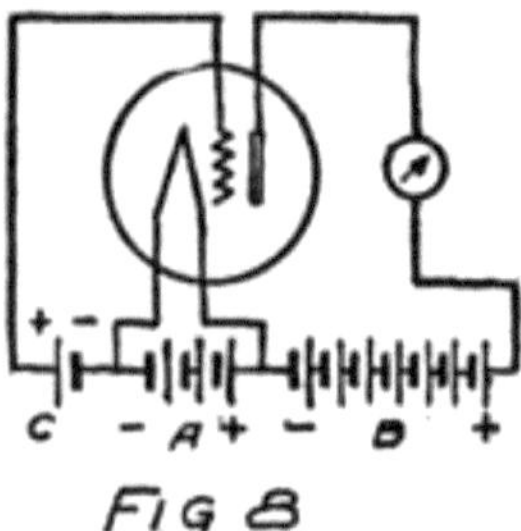

FIG 8

What happens then is something like this: Electrons are freed at the filament; the plate and the grid both call them and they start off in a rush. Some of them are stopped by the wires of the grid but most of them go on by to the plate. The grid is mostly open space, you know, and the electrons move as fast as lightning. They are going too fast in the general direction of the grid to stop and look for its few and small wires.

When the grid is positive the grid helps the plate to call electrons away from the filament. Making the grid positive, therefore, increases the stream of electrons *between filament and plate*; that is, increases the current in the plate circuit.

We could get the same effect so far as concerns 48an increased plate current by using more batteries in series in the plate circuit so as to pull harder. But the grid is so close to the filament that a single battery cell in the grid circuit can call electrons so strongly that it would take several extra battery cells in the plate circuit to produce the same effect.

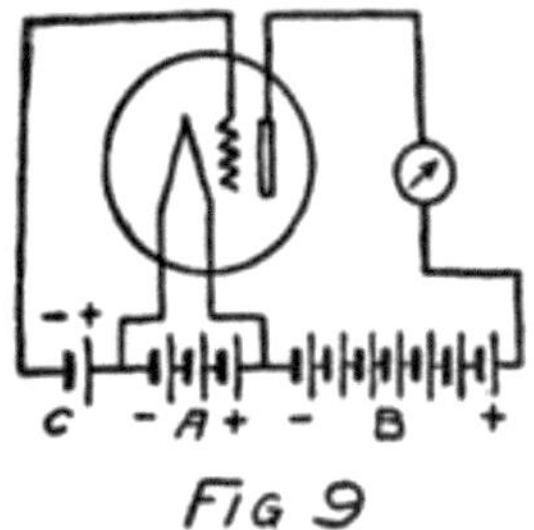

FIG 9

If we reverse the grid battery, as in Fig. 9, so as to make the grid negative, then, instead of attracting electrons the grid repels them. Nowhere near as many electrons will stream across to the plate when the grid says, "No, go back." The grid is in a strategic position and what it says has a great effect.

When there is no battery connected to the grid it has no possibility of influencing the electron stream which the plate is attracting to itself. We say, then, that the grid is uncharged or is at "zero potential," meaning that it is zero or nothing in possibility. But when the grid is charged, no matter how little, it makes a change in the plate current. When the grid says "Come on," even though very softly, it has as much effect on the electrons as if the plate shouted at them, and a lot of extra electrons rush for the plate. But when the grid whispers "Go back," many electrons which would otherwise have gone streaking off to the plate crowd back toward the filament. That's how the audion works. There is an electron stream and a wonderfully sensitive way of controlling the stream.

49 LETTER 7
HOW TO MEASURE AN ELECTRON STREAM

(This letter may be omitted on the first reading.)

DEAR YOUTH:

If we are to talk about the audion and how its grid controls the current in the plate circuit we must know something of how to measure currents. An electric current is a stream of electrons. We measure it by finding the rate at which electrons are traveling along through the circuit.

What do we mean by the word "rate?" You know what it means when a speedometer says twenty miles an hour. If the car should keep going just as it was doing at the instant you looked at the speedometer it would go twenty miles in the next hour. Its rate is twenty miles an hour even though it runs into a smash the next minute and never goes anywhere again except to the junk heap.

It's the same when we talk of electric currents. We say there is a current of such and such a number of electrons a second going by each point in the circuit. We don't mean that the current isn't going to change, for it may get larger or smaller, but we do mean that if the stream of electrons keeps going just as it is there will be such and such a number of electrons pass by in the next second.

In most of the electrical circuits with which you 50will deal you will find that electrons must be passing along in the circuit at a most amazing rate if there is to be any appreciable effect. When you turn on the 40-watt light at your desk you start them going through the filament of the lamp at the rate of about two and a half billion billion each second. You have stood on the sidewalk in the city and watched the people stream past you. Just suppose you could stand beside that narrow little sidewalk which the filament offers to the electrons and count them as they go by. We don't try to count them although we do to-day know about how many go by in a second if the current is steady.

If some one asks you how old you are you don't say "About five hundred million seconds"; you tell him in years. When some one asks how large a current is flowing in a wire we don't tell him six billion billion electrons each second; we tell him "one ampere." Just as we use years as the units in which to count up time so we use amperes as the units in which to count up streams of electrons. When a wire is carrying a current of one ampere the electrons are streaming through it at the rate of about 6,000,000,000,000,000,000 a second.

Don't try to remember this number but do remember that an ampere is a unit in which we measure currents just as a year is a unit in which we measure time. An ampere is a unit in which we measure streams of electrons just as "miles per hour" is a unit in which we measure the speed of trains or automobiles.

51If you wanted to find the weight of something you would take a scale and weigh it, wouldn't you? You might take that spring balance which hangs out in the kitchen. But if the spring balance said the thing weighed five pounds how would you know if it was right? Of course you might take what ever it was down town and weigh it on some other scales but how would you know those scales gave correct weight?

The only way to find out would be to try the scales with weights which you were sure were right and see if the readings on the scale correspond to the known weights. Then you could trust it to tell you the weight of something else. That's the way scales are tested. In fact that's the way that the makers know how to mark them in the first place. They put on known weights and marked the lines and figures which you see. What they did was called "calibrating" the scale. You could make a scale for yourself if you wished, but if it was to be reliable you would have to find the places for the markings by applying known weights, that is, by calibration.

How would you know that the weights you used to calibrate your scale were really what you thought them to be? You would have to find some place where they had a weight that everybody would agree was correct and then compare your weight with that. You might, for example, send your pound weight to the Bureau of Standards in Washington and for a small payment have the Bureau compare it with the pound which it keeps as a standard.

52That is easy where one is interested in a pound. But it is a little different when one is interested in an ampere. You can't make an ampere out of a piece of platinum as you can a standard pound weight. An ampere is a stream of electrons at about the rate of six billion billion a second. No one could ever count anywhere near that many, and yet everybody who is concerned with electricity wants to be able to measure currents in amperes. How is it done?

First there is made an instrument which will have something in it to move when electrons are flowing through the instrument. We want a meter for the flow of electrons. In the basement we have a meter for the flow of gas and another for the flow of water. Each of these has some part which will move when the water or the gas passes through. But they are both arranged with little gear wheels so as to keep track of all the water or gas which has flowed through; they won't tell the rate at which the gas or water is flowing. They are like the odometer on the car which gives the "trip mileage" or the "total mileage." We want a meter like the speedometer which will indicate at each instant just how fast the electrons are streaming through it.

There are several kinds of meters but I shall not try to tell you now of more than one. The simplest to understand is called a "hot-wire meter." You already know that an electron stream heats a wire. Suppose a piece of fine wire is fastened at the two ends and that there are binding posts also fastened 53to these ends of the wire so that the wire may be made part of the circuit where we want to know the electron stream. Then the same stream of electrons will flow through the fine wire as through the other parts of the circuit. Because the wire is fine it acts like a very narrow sidewalk for the stream of electrons and they have to bump and jostle pretty hard to get through. That's why the wire gets heated.

You know that a heated wire expands. This wire expands. It grows longer and because it is held firmly at the ends it must bow out at the center. The bigger the rate of flow of electrons the hotter it gets; and the hotter it gets the more it bows out. At the center we might fasten one end–the short end–of a little lever. A small motion of this short end of the lever will mean a large motion of the other end, just like a "teeter board" when one end is longer than the other;

the child on the long end travels further than the child on the short end. The lever magnifies the motion of the center of the hot wire part of our meter so that we can see it easier.

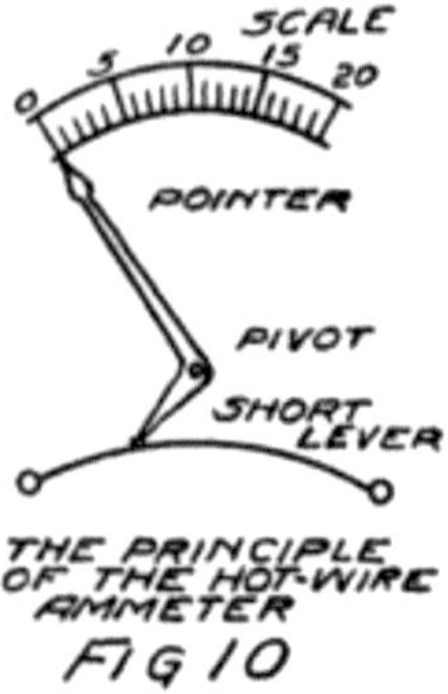

There are several ways to make such a meter. The one shown in Fig. 10 is as easy to understand as any. We shape the long end of the lever like a pointer. Then the hotter the wire the farther the pointer moves.

If we could put this meter in an electric circuit 54where we knew one ampere was flowing we could put a numeral "1" opposite where the pointer stood. Then if we could increase the current until there were two amperes flowing through the meter we could mark that position of the pointer "2" and so on. That's the way we would calibrate the meter. After we had done so we would call it an "ammeter" because it measures amperes. Years ago people would have called it an "amperemeter" but no one who is up-to-date would call it so to-day.

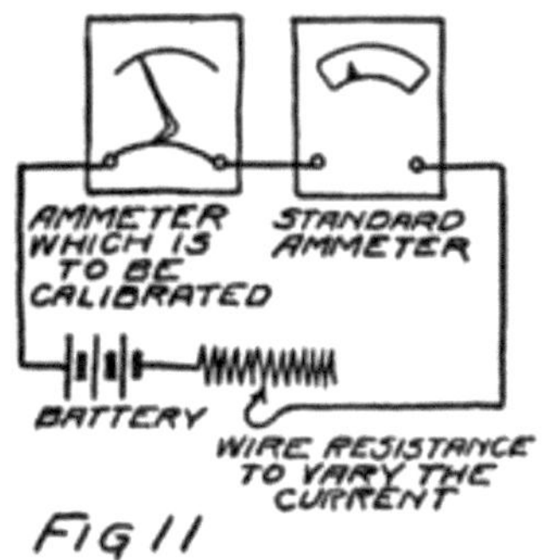

FIG 11

If we had a very carefully made ammeter we would send it to the Bureau of Standards to be calibrated. At the Bureau they have a number of meters which they know are correct in their readings. They would put one of their meters and ours into the same circuit so that both carry the same stream of electrons as in Fig. 11. Then whatever the reading was on their meter could be marked opposite the pointer on ours.

Now I want to tell you how the physicists at the Bureau know what is an ampere. Several years ago there was a meeting or congress of physicists and electrical engineers from all over the world who discussed what they thought should be the unit in which to measure current. They decided just what they would call an ampere and then all the countries from which they came passed laws saying that an ampere 55should be what these scientists had recommended. To-day, therefore, an ampere is defined by law.

To tell when an ampere of current is flowing requires the use of two silver plates and a solution of silver nitrate. Silver nitrate has molecules made up of one atom of silver combined with a group of atoms called "nitrate." You remember that the molecule of copper sulphate, discussed in our third letter, was formed by a copper atom and a group called sulphate. Nitrate is another group something like sulphate for it has oxygen atoms in it, but it has three instead of four, and instead of a sulphur atom there is an atom of nitrogen.

When silver nitrate molecules go into solution they break up into ions just as copper sulphate does. One ion is a silver atom which has lost one electron. This electron was stolen from it by the nitrate part of the molecule when they dissociated. The nitrate ion, therefore, is formed by a nitrogen atom, three oxygen atoms, and one extra electron.

If we put two plates of silver into such a solution nothing will happen until we connect a battery to the plates. Then the battery takes electrons away from one plate and gives electrons to the other. Some of the atoms in the plate which the battery is robbing of electrons are just like the silver ions which are moving around in the solution. That's why they can go out into the solution and play with the nitrate ions each of which has an extra electron which it stole from some silver atom. But the moment silver ions 56leave their plate we have more silver ions in the solution than we do sulphate ions.

The only thing that can happen is for some of the silver ions to get out of the solution. They aren't going back to the positive silver plate from which they just came. They go on toward the negative plate where the battery is sending an electron for every one which it takes away from the positive plate. There start off towards the negative plate, not only the ions which just came from the positive plate, but all the ions that are in the solution. The first one to arrive gets an electron but it can't take it away from the silver plate. And why should it? As soon as it has got this electron it is again a normal silver atom. So it stays with the other atoms in the silver plate. That's what happens right along. For every atom which is lost from the positive plate there is one added to the negative plate. The silver of the positive plate gradually wastes away and the negative plate gradually gets an extra coating of silver.

Every time the battery takes an electron away from the positive plate and gives it to the negative plate there is added to the negative plate an atom of silver. If the negative plate is weighed before the battery is connected and again after the battery is disconnected we can tell how much silver has been added to it. Suppose the current has been perfectly steady, that is, the same number of electrons streaming through the circuit each second. Then if we 57know how long the current has been running we can tell how much silver has been deposited each second.

The law says that if silver is being deposited at the rate of 0.001118 gram each second then the current is one ampere. That's a small amount of silver, only about a thousandth part of a gram, and you know that it takes 28.35 grams to make an ounce. It's a very small amount of silver but it's an enormous number of atoms. How many? Six billion billion, of course, for there is deposited one atom for each electron in the stream.

In my next letter I'll tell you how we measure the pull which batteries can give to electrons, and then we shall be ready to go on with more about the audion.

58 LETTER 8
ELECTRON-MOVING-FORCES

(This letter may be omitted on the first reading.)

DEAR YOUNG MAN:

I trust you have a fairly good idea that an ampere means a stream of electrons at a certain definite rate and hence that a current of say 3 amperes means a stream with three times as many electrons passing along each second.

In the third and fourth letters you found out why a battery drives electrons around a conducting circuit. You also found that there are several different kinds of batteries. Batteries differ in their abilities to drive electrons and it is therefore convenient to have some way of comparing them. We do this by measuring the electron-moving-force or "electromotive force" which each battery can exert. To express electromotive force and give the results of our measurements we must have some unit. The unit we use is called the "volt."

The volt is defined by law and is based on the suggestions of the same body of scientists who recommended the ampere of our last letter. They defined it by telling how to make a particular kind of battery and then saying that this battery had an electromotive force of a certain number of volts. One can buy such standard batteries, or standard 59cells as they are called, or he can make them for himself. To be sure that they are just right he can then send them to the Bureau of Standards and have them compared with the standard cells which the Bureau has.

I don't propose to tell you much about standard cells for you won't have to use them until you come to study physics in real earnest. They are not good for ordinary purposes because the moment they go to work driving electrons the conditions inside them change so their electromotive force is changed. They are delicate little affairs and are useful only as standards with which to compare other batteries. And even as standard batteries they must be used in such a way that they are not required to drive any electrons.

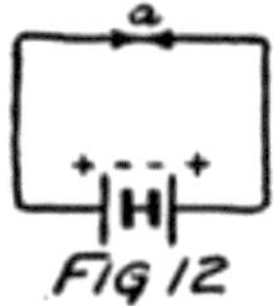

Let's see how it can be done. Suppose two boys sit opposite each other on the floor and brace their feet together. Then with their hands they take hold of a stick and pull in opposite directions. If both have the same stick-motive-force the stick will not move.

Now suppose we connect the negative feet–I mean negative terminals–of two batteries together as in Fig. 12. Then we connect their positive terminals together by a wire. In the wire there will be lots of free electrons ready to go to the positive plate of the battery which pulls the harder. If the batteries are equal in electromotive force these electrons will stay right where they are. There will be no stream 60of electrons and yet we'll be using one of the batteries to compare with the other.

That is all right, you think, but what are we to do when the batteries are not just equal in e. m. f.? (e. m. f. is code for electromotive force). I'll tell you, because the telling includes some other ideas which will be valuable in your later reading.

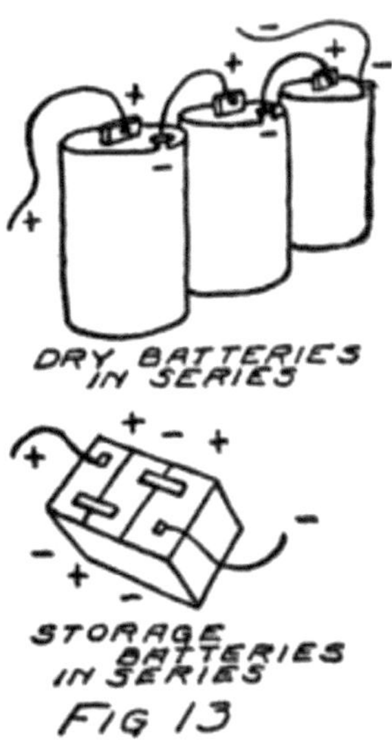

Suppose we take batteries which aren't going to be injured by be-ing made to work–storage batteries will do nicely–and connect them in series as in Fig. 13. When batteries are in series they act like a single stronger battery, one whose e. m. f. is the sum of the e. m. f.'s of the separate batteries. Connect these batteries to a long fine wire as in Fig. 14. There is a stream of electrons along this wire. Next connect the negative terminal of the standard cell to the negative terminal of the storage batteries, that is, brace their feet against each other. Then connect a wire to the positive terminal of the standard cell. This wire acts just like a long arm sticking out from the positive plate of this cell.

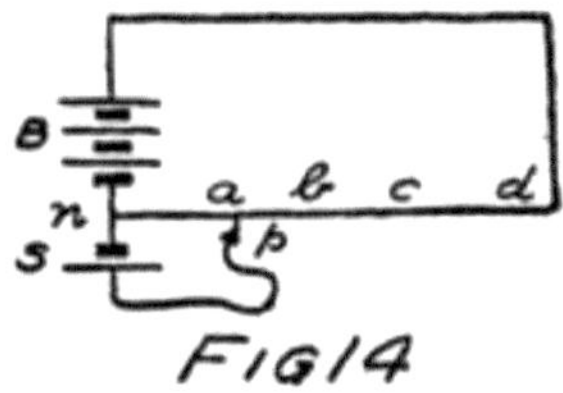

Touch the end of the wire, which is p of Fig. 14, 61 to some point as a on the fine wire. Now what do we have? Right at a, of course, there are some free electrons and they hear the calls of both batteries. If the standard battery, *S* of the figure, calls the stronger they go to it. In that case move the end *p* nearer the positive plate of the battery *B*, so that it will have a chance to exert a stronger pull. Suppose we try at *c* and find the battery *B* is there the stronger. Then we can move back to some point, say *b*, where the pulls are equal.

To make a test like this we put a sensitive current-measuring instrument in the wire which leads from the positive terminal of the standard cell. We also use a long fine wire so that there can never be much of an electron stream anyway. When the pulls are equal there will be no current through this instrument.

As soon as we find out where the proper setting is we can replace *S* by some other battery, say *X*, which we wish to compare with *S*. We find the setting for that battery in the same way as we just did for *S*. Suppose it is at *d* in Fig. 14 while the setting for *S* was at *b*. We can see at once that *X* is stronger than *S*. The question, however, is how much stronger.

Perhaps it would be better to try to answer this question by talking about e. m. f.'s. It isn't fair to speak only of the positive plate which calls, we must speak also of the negative plate which is shooing electrons away from itself. The idea of e. m. f. takes care of both these actions. The steady stream of 62electrons in the fine wire is due to the e. m. f. of the battery *B*, that is to the pull of the positive terminal and the shove of the negative.

If the wire is uniform, that is the same throughout its length, then each inch of it requires just as much e. m. f. as any other inch. Two inches require twice the e. m. f. which one inch requires. We know how much e. m. f. it takes to keep the electron stream going in the part of the wire from *n* to *b*. It takes just the e. m. f. of the standard cell, *S*, because when that had its feet braced at *n* it pulled just as hard at *b* as did the big battery *B*.

Suppose the distance *n* to *d* (usually written *nd*) is twice as great as that from *n* to *b* (*nb*). That means that battery *X* has twice the e. m. f. of battery *S*. You remember that *X* could exert the same force through the length of wire *nd*, as could the large battery. That is twice what cell *S* can do. Therefore if we know how many volts to call the e. m. f. of the standard cell we can say that *X* has an e. m. f. of twice as many volts.

If we measured dry batteries this way we should find that they each had an e. m. f. of about 1.46 volts. A storage battery would be found to have about 2.4 volts when fully charged and perhaps as low as 2.1 volts when we had run it for a while.

That is the way in which to compare batteries and to measure their e. m. f.'s, but you see it takes a lot of time. It is easier to use a "voltmeter" which is an instrument for measuring e. m. f.'s. Here is how one could be made.

63First there is made a current-measuring instrument which is quite sensitive, so that its pointer will show a deflection when only a very small stream of electrons is passing through the instrument. We could make one in the same way as we made the ammeter of the last letter but there are other better ways of which I'll tell you later. Then we connect a good deal of fine wire in series with the instrument for a reason which I'll tell you in a minute. The next and last step is to calibrate.

We know how many volts of e. m. f. are required to keep going the electron stream between n and b–we know that from the e. m. f. of our standard cell. Suppose then that we connect this new instrument, which we have just made, to the wire at n and b as in Fig. 15. Some of the electrons at n which are so anxious to get away from the negative plate of battery B can now travel as far as b through the wire of the new instrument. They do so and the pointer swings around to some new position. Opposite that we mark the number of volts which the standard battery told us there was between n and b.

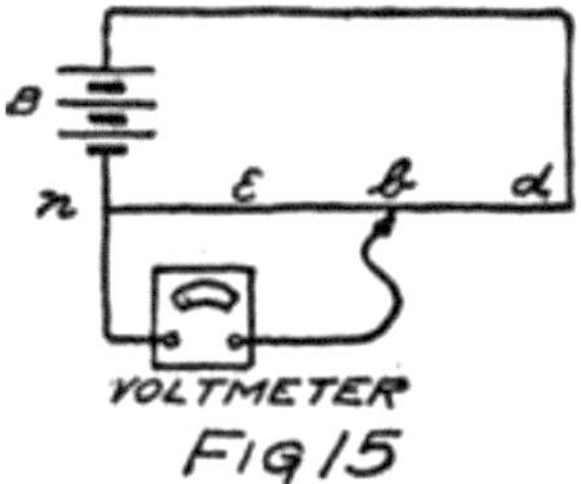

If we move the end of the wire from b to d the pointer will take a new position. Opposite this we mark twice the number of volts of the standard cell. We can run it to a point e where the distance ne is one-half nb, and mark our scale with half the number of volts of the standard cell, and so on for other 64 positions along the wire. That's the way we calibrate a sensitive current-measuring instrument (with its added wire, of course) so that it will read volts. It is now a voltmeter.

If we connect a voltmeter to the battery X as in Fig. 16 the pointer will tell us the number of volts in the e. m. f. of X, for the pointer will take the same position as it did when the voltmeter was connected between n and d.

There is only one thing to watch out for in all this. We must be careful that the voltmeter is so made that it won't offer too easy a path for electrons to follow. We only want to find how hard a battery can pull an electron, for that is what we mean by e. m. f. Of course, we must let a small stream of electrons flow through the voltmeter so as to make the pointer move. That is why voltmeters of this kind are made out of a long piece of fine wire or else have a coil of fine wire in series with the current-measuring part. The fine wire makes a long and narrow path for the electrons and so there can be only a small stream. Usually we describe this condition by saying that a voltmeter has a high resistance.

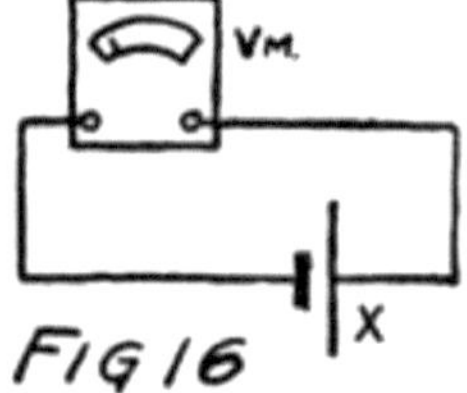

Fine wires offer more resistance to electron streams than do heavy wires of the same length. If a wire is the same diameter all along, the longer the length of it which we use the greater is the resistance which is offered to an electron stream.

65You will need to know how to describe the resistance of a wire or of any part of an electric circuit. To do so you tell how many "ohms" of resistance it has. The ohm is the unit in which we measure the resistance of a circuit to an electron stream.

I can show you what an ohm is if I tell you a simple way to measure a resistance. Suppose you have a wire or coil of wire and want to know its resistance. Connect it in series with a battery and an ammeter as shown in Fig. 17. The same electron stream passes through all parts of this circuit and the ammeter tells us what this stream is in amperes. Now connect a voltmeter to the two ends of the coil as shown in the figure. The voltmeter tells in volts how much e. m. f. is being applied to force the current through the coil. Divide the number of volts by the number of amperes and the quotient (answer) is the number of ohms of resistance in the coil.

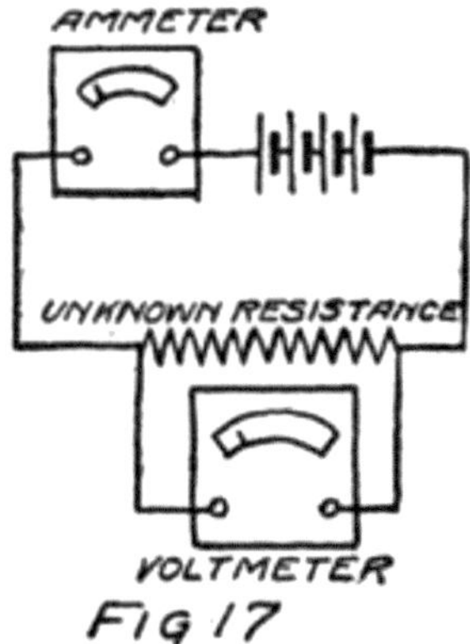

Suppose the ammeter shows a current of one ampere and the voltmeter an e. m. f. of one volt. Then dividing 1 by 1 gives 1. That means that the coil has a resistance of one ohm. It also means one ohm is such a resistance that one volt will send through it a current of one ampere. You can get lots of meaning out of this. For example, it means also 66that one volt will send a current of one ampere through a resistance of one ohm.

How many ohms would the coil have if it took 5 volts to send 2 amperes through it. Solution: Divide 5 by 2 and you get 2.5. Therefore the coil would have a resistance of 2.5 ohms.

Try another. If a coil of resistance three ohms is carrying two amperes what is the voltage across the terminals of the coil? For 1 ohm it would take 1 volt to give a current of 1 ampere, wouldn't it? For 3 ohms it takes three times as much to give one ampere. To give twice this current would take twice 3 volts. That is, 2 amperes in 3 ohms requires 2x3 volts.

Here's one for you to try by yourself. If an e. m. f. of 8 volts is sending current through a resistance of 2 ohms, how much current is flowing? Notice that I told the number of ohms and the number of volts, what are you going to tell? Don't tell just the number; tell how many and what.

67 LETTER 9
THE AUDION-CHARACTERISTIC

Although there is much in Letters 7 and 8 which it is well to learn and to think about, there are only three of the ideas which you must have firmly grasped to get the most out of this letter which I am now going to write you about the audion.

First: Electric currents are streams of electrons. We measure currents in amperes. To measure a current we may connect into the circuit an ammeter.

Second: Electrons move in a circuit when there is an electron-moving-force, that is an electromotive force or e. m. f. We measure e. m. f.'s in volts. To measure an e. m. f. we connect a voltmeter to the two points between which the e. m. f. is active.

Third: What current any particular e. m. f. will cause depends upon the circuit in which it is active. Circuits differ in the resistance which they offer to e. m. f.'s. For any particular e. m. f. (that is for any given e. m. f.) the resulting current will be smaller the greater the resistance of the circuit. We measure resistance in ohms. To measure it we find the quotient of the number of volts applied to the circuit by the number of amperes which flow.

In my sixth letter I told you something of how the audion works. It would be worth while to read again that letter. You remember that the current in the 68plate circuit can be controlled by the e. m. f. which is applied to the grid circuit. There is a relationship between the plate current and the grid voltage which is peculiar or characteristic to the tube. So we call such a relationship "a characteristic." Let us see how it may be found and what it will be.

Connect an ammeter in the plate- or B-circuit, of the tube so as to measure the plate-circuit current. You will find that almost all books use the letter "I" to stand for current. The reason is that scientists used to speak of the "intensity of an electric current" so that "I" really stands for intensity. We use I to stand for something more than the word "current." It is our symbol for whatever an ammeter would read, that is for the amount of current.

Another convenience in symbols is this: We shall frequently want to speak of the currents in several different circuits. It saves time to use another letter along with the letter I to show the circuit to which we refer. For example, we are going to talk about the current in the B-circuit of the audion, so we call that current I_B . We write the letter B below the line on which I stands. That is why we say the B is subscript, meaning "written below." When you are reading to yourself be sure to read I_B as "eye-bee" or else as "eye-subscript-bee." I_B therefore will stand for the number of amperes in the 69 plate circuit of the audion. In the same way I_A would stand for the current in the filament circuit.

We are going to talk about e. m. f.'s also. The letter "E" stands for the number of volts of e. m. f. in a circuit. In the filament circuit the battery has E_A volts. In the plate circuit the e. m. f. is E_B volts. If we put a battery in the grid circuit we can let E_C represent the number of volts applied to the grid-filament or C-circuit.

The characteristic relation which we are after is one between grid voltage, that is E_C , and plate current, that is I_B . So we call it the E_C – I_B characteristic. The dash between the letters is not a subtraction sign but merely a dash to separate the letters. Now we'll find the "ee-see-eye-bee" characteristic.

Connect some small dry cells in series for use in the grid circuit. Then connect the filament to the middle cell as in Fig. 19. Take the wire which comes from the grid and put a battery clip on it, then you can connect the grid anywhere you want along this series of batteries. See Fig. 18. In the figure this movable clip is represented by an arrow head. You can see that if it is at a the battery will make the grid positive. If it is moved to b the grid will be more positive. On the other hand if the clip is at o there will be no e. m. f. applied to the grid. If it is at c the grid will be made negative.

Between grid and filament there is placed a voltmeter which will tell how much e. m. f. is applied to the grid, that is, tell the value of E_C, for any position whatever of the clip.

70We shall start with the filament heated to a deep red. The manufacturers of the audion tell the purchaser what current should flow through the filament so that there will be the proper emission of electrons. There are easy ways of finding out for one's self but we shall not stop to describe them. The makers also tell how many volts to apply to the plate, that is what value E_B should have. We could find this out also for ourselves but we shall not stop to do so.

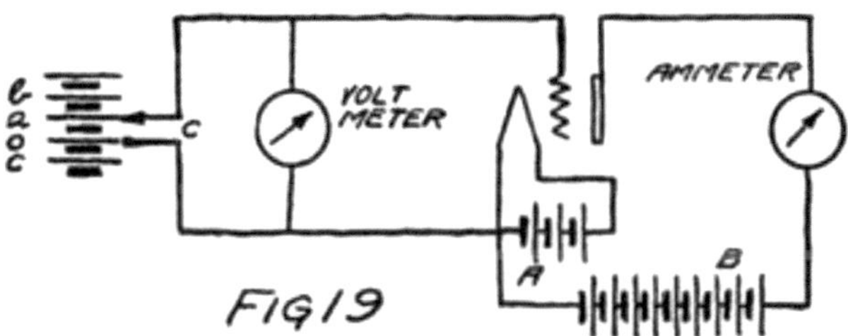

Now we set the battery clip so that there is no voltage applied to the grid; that is, we start with E_C equal to zero. Then we read the ammeter in the plate circuit to find the value of I_B which corresponds to this condition of the grid.

Next we move the clip so as to make the grid as positive as one battery will make it, that is we move the clip to a in Fig. 19. We now have a different value of E_C and will find a different value of I_B when we read the ammeter. Next move the clip to apply two batteries to the grid. We get a new pair of values for E_C and I_B, getting E_C from the voltmeter and I_B from the ammeter. As we continue in this way, increasing E_C, we find that the current I_B increases 71 for a while and then after we have reached a certain value of E_C the current I_B stops increasing. Adding more batteries and making the grid more positive doesn't have any effect on the plate current.

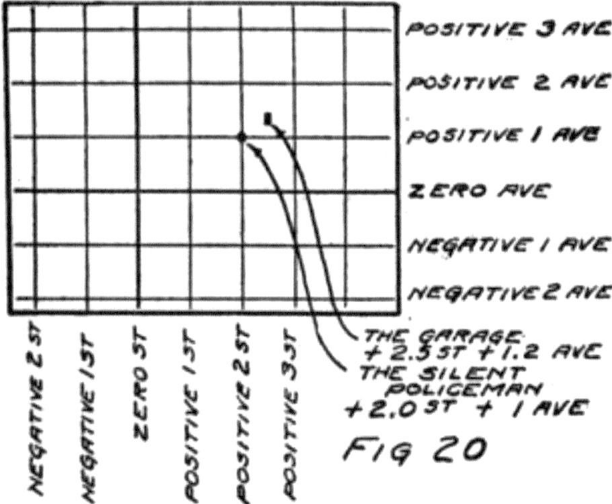

FIG 20

Before I tell you why this happens I want to show you how to make a picture of the pairs of values of E_C and I_B which we have been reading on the voltmeter and ammeter.

Imagine a city where all the streets are at right angles and the north and south streets are called streets and numbered while the east and west thorofares are called avenues. I'll draw the map as in Fig. 20. Right through the center of the city goes Main Street. But the people who laid out the roads were mathematicians and instead of calling it Main Street they called it "Zero Street." The first street east of Zero St. we should have called "East First Street" but they called it "Positive 1 St." and the 72next beyond "Positive 2 St.," and so on. West of the main street they called the first street "Negative 1 St." and so on.

When they came to name the avenues they were just as precise and mathematical. They called the main avenue "Zero Ave." and those north of it "Positive 1 Ave.," "Positive 2 Ave." and so on. Of course, the avenues south of Zero Ave. they called Negative.

The Town Council went almost crazy on the subject of numbering; they numbered everything. The silent policeman which stood at the corner of "Positive 2 St." and "Positive 1 Ave." was marked that way. Half way between Positive 2 St. and Positive 3 St. there was a garage which set back about two-tenths of a block from Positive 1 Ave. The Council numbered it and called it "Positive 2.5 St. and Positive 1.2 Ave." Most of the people spoke of it as "Plus 2.5 St. and Plus 1.2 Ave."

Sometime later there was an election in the city and a new Council was elected. The members were mostly young electricians and the new Highway Commissioner was a radio enthusiast. At the first meeting the Council changed the names of all the avenues to "Mil-amperes" [3] and of all the streets to "Volts."

Then the Highway Commissioner who had just been taking a set of voltmeter and ammeter readings on an audion moved that there should be a new 73road known as "Audion Characteristic." He said the road should pass through the following points:

Zero Volt and Plus 1.0 Mil-ampere
Plus 2.0 Volts and Plus 1.7 Mil-amperes
Plus 4.0 Volts and Plus 2.6 Mil-amperes
Plus 6.0 Volts and Plus 3.4 Mil-amperes
Plus 8.0 Volts and Plus 4.3 Mil-amperes

And so on. Fig. 21 shows the new road.

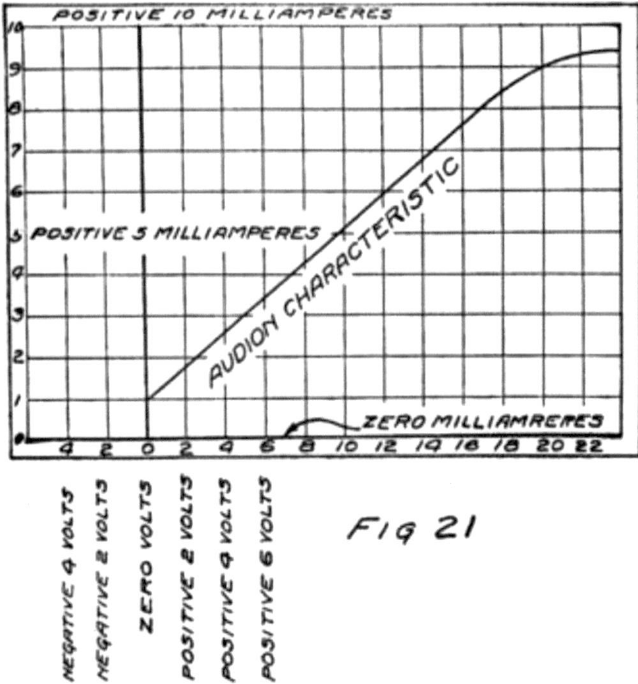

FIG 21

One member of the Council jumped up and said "But what if the grid is made negative?" The Commissioner had forgotten to see what happened so he went home to take more readings.

He shifted the battery clip along, starting at c of 74 Fig. 22. At the next meeting of the Council he brought in the following list of readings and hence of points on his proposed road.

Minus	1.0	Volts	and	Plus	0.6	Mil-ampere
"	2.0	"	"	"	0.4	"
"	3.0	"	"	"	0.2	"
"	4.0	"	"	"	0.1	"
"	5.0	"	"	"	0.0	"

Then he showed the other members of the Council on the map of Fig. 23 how the Audion Characteristic would look.

74

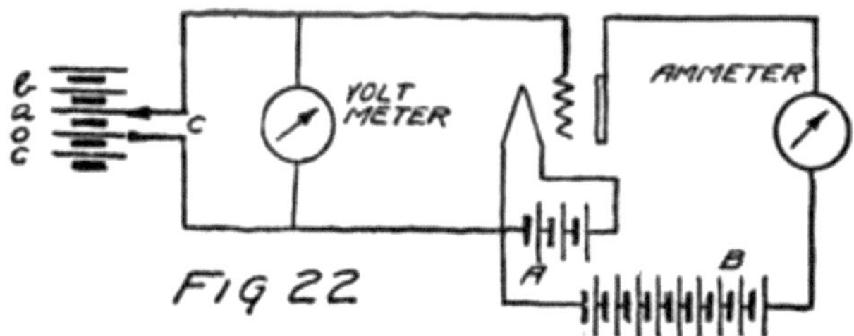

There was considerable discussion after that and it appeared that different designs and makes of audions would have different characteristic curves. They all had the same general form of curve but they would pass through different sets of points depending upon the design and upon the B-battery voltage. It was several meetings later, however, before they found out what effects were due to the form of the curve. Right after this they found that they could get much better results with their radio sets.

Now look at the audion characteristic. Making the grid positive, that is going on the positive side of the zero volts in our map, makes the plate current 75larger. You remember that I told you in Letter 6 how the grid, when positive, helped call electrons away from the filament and so made a larger stream of electrons in the plate circuit. The grid calls electrons away from the filament. It can't call them out of it; they have to come out themselves as I explained to you in the fifth letter.

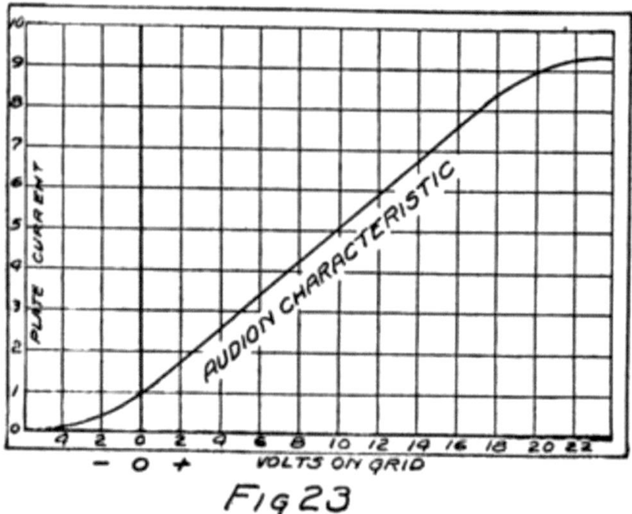

FIG 23

You can see that as we make the grid more and more positive, that is, make it call louder and louder, a condition will be reached where it won't do it any good to call any louder, for it will already be getting all the electrons away from the filament just as fast as they are emitted. Making the grid more positive after that will not increase the plate current any. That's why the characteristic flattens off as you see at high values of grid voltage.

The arrangement which we pictured in Fig. 22 for 76making changes in the grid voltage is simple but it doesn't let us change the voltage by less than that of a single battery cell. I want to show you a way which will. You'll find it very useful to know and it is easily understood for it is something like the arrangement of Fig. 14 in the preceding letter.

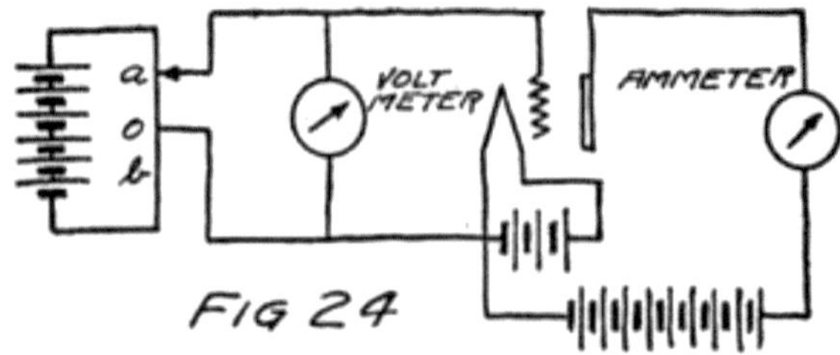

FIG 24

Connect the cells as in Fig. 24 to a fine wire. About the middle of this wire connect the filament. As before use a clip on the end of the wire from the grid. If the grid is connected to *a* in the figure there is applied to the grid circuit that part of the e. m. f. of the battery which is active in the length of wire between *o* and *a*. The point *a* is nearer the positive plate of the battery than is the point *o*. So the grid will be positive and the filament negative.

On the other hand, if the clip is connected at *b* the grid will be negative with respect to the filament. We can, therefore, make the grid positive or negative depending on which side of *o* we connect the clip. How large the e. m. f. is which will be applied to the grid depends, of course, upon how far away from *o* the clip is connected.

Suppose you took the clip in your hand and slid it along in contact with the wire, first from *o* to *a* 77 and then back again through *o* to *b* and so on back and forth. You would be making the grid *alternately* positive and negative, wouldn't you? That is, you would be applying to the grid an e. m. f. which increases to some positive value and then, decreasing to zero, *reverses*, and increases just as much, only to decrease to zero, where it started. If you do this over and over again, taking always the same time for one round trip of the clip you will be impressing on the grid circuit an "*alternating e. m. f.*"

What's going to happen in the plate circuit? When there is no e. m. f. applied to the grid circuit, that is when the grid potential (possibilities) is zero, there is a definite current in the plate circuit. That current we can find from our characteristic of Fig. 23 for it is where the curve crosses Zero Volts. As the grid becomes positive the current rises above this value. When the grid is made negative the current falls below this value. The current, I_B , then is made alternately greater and less than the current when E_C is zero.

You might spend a little time thinking over this, seeing what happens when an alternating e. m. f. is applied to the grid of an audion, for that is going to be fundamental to our study of radio.

[3]

A mil-ampere is a thousandth of an ampere just as a millimeter is a thousandth of a meter.

 LETTER 10
CONDENSERS AND COILS

DEAR SON:

In the last letter we learned of an alternating e. m. f. The way of producing it, which I described, is very crude and I want to tell how to make the audion develop an alternating e. m. f. for itself. That is what the audion does in the transmitting set of a radio telephone. But an audion can't do it all alone. It must have associated with it some coils and a condenser. You know what I mean by coils but you have yet to learn about condensers.

A condenser is merely a gap in an otherwise conducting circuit. It's a gap across which electrons cannot pass so that if there is an e. m. f. in the circuit, electrons will be very plentiful on one side of the gap and scarce on the other side. If there are to be many electrons waiting beside the gap there must be room for them. For that reason we usually provide waiting-rooms for the electrons on each side of the gap. Metal plates or sheets of tinfoil serve nicely for this purpose. Look at Fig. 25. You see a battery and a circuit which would be conducting except for the gap at C. On each side of the gap there is a sheet of metal. The metal sheets may be separated by air or mica or paraffined paper. The 79combination of gap, plates, and whatever is between, provided it is not conducting, is called a condenser.

Let us see what happens when we connect a battery to a condenser as in the figure. The positive terminal of the battery calls electrons from one plate of the condenser while the negative battery-terminal drives electrons away from itself toward the other plate of the condenser. One plate of the condenser, therefore, becomes positive while the other plate becomes negative.

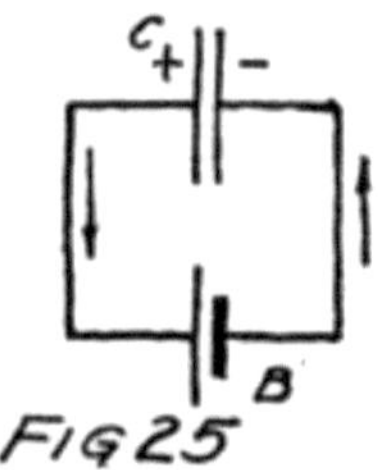

You know that this action of the battery will go on until there are so many electrons in the negative plate of the condenser that they prevent the battery from adding any more electrons to that plate. The same thing happens at the other condenser plate. The positive terminal of the battery calls electrons away from the condenser plate which it is making positive until so many electrons have left that the protons in the atoms of the plate are calling for electrons to stay home just as loudly and effectively as the positive battery-terminal is calling them away.

When both these conditions are reached–and they are both reached at the same time–then the battery has to stop driving electrons around the circuit. The battery has not enough e. m. f. to drive any more electrons. Why? Because the condenser has now just enough e. m. f. with which to oppose the battery.

It would be well to learn at once the right words 80to use in describing this action. We say that the battery sends a "charging current" around its circuit and "charges the condenser" until it has the same e. m. f. When the battery is first connected to the condenser there is lots of space in the waiting-rooms so there is a great rush or surge of electrons into one plate and away from the other. Just at this first instant the charging current, therefore, is large but it decreases rapidly, for the moment electrons start to pile up on one plate of the condenser and to leave the other, an e. m. f. builds up on the condenser. This e. m. f., of course, opposes that of the battery so that the net e. m. f. acting to move electrons round the circuit is no longer that of the battery, but is the difference between the e. m. f. of the battery and that of the condenser. And so, with each added electron, the e. m. f. of the condenser increases until finally it is just equal to that of the battery and there is no net e. m. f. to act.

What would happen if we should then disconnect the battery? The condenser would be left with its extra electrons in the negative plate and with its positive plate lacking the same number of electrons. That is, the condenser would be left charged and its e. m. f. would be of the same number of volts as the battery.

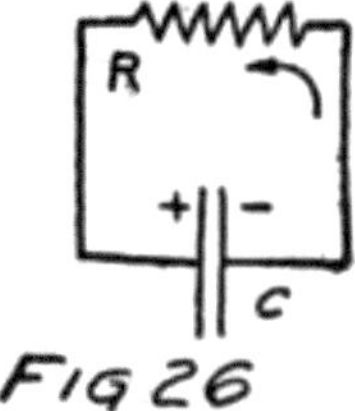

FIG 26

Now suppose we connect a short wire between the plates of the condenser as in Fig. 26. The electrons rush home from the negative plate to the positive plate. As fast as electrons get home 81the e. m. f. decreases. When they are all back the e. m. f. has been reduced to zero. Sometimes we say that "the condenser discharges." The "discharge current" starts with a rush the moment the conducting path is offered between the two plates. The e. m. f. of the condenser falls, the discharge current grows smaller, and in a very short time the condenser is completely discharged.

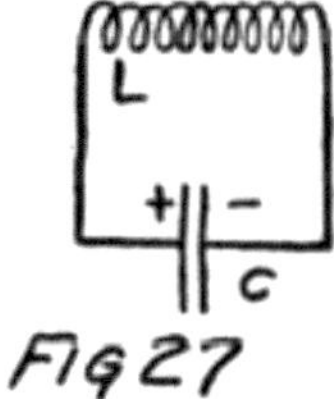

FIG 27

That's what happens when there is a short conducting path for the discharge current. If that were all that could happen I doubt if there would be any radio communication to-day. But if we connect a coil of wire between two plates of a charged condenser, as in Fig. 27, then something of great interest happens. To understand you must know something more about electron streams.

Suppose we should wind a few turns of wire on a cylindrical core, say on a stiff cardboard tube. We shall use insulated wire. Now start from one end of the coil, say *a*, and follow along the coiled wire for a few turns and then scratch off the insulation and solder onto the coil two wires, *b*, and *c*, as shown in Fig. 28. The further end of the coil we shall call *d*. Now let's arrange a battery and switch so that we can send a current through the part of the coil between *a* and *b*. Arrange also a current-measuring instrument so as to show if any current is flowing in the part of the coil between *c* and *d*. For this purpose we shall use a kind of current-measuring instrument which I have not yet explained. It is different from the hot-wire type described in Letter 7 for it will show in which direction electrons are streaming through it.

The diagram of Fig. 28 indicates the apparatus of our experiment. When we close the switch, *S*, the battery starts a stream of electrons from *a* towards *b*. Just at that instant the needle, or pointer, of the current instrument moves. The needle moves, and thus shows a current in the coil *cd*; but it comes right back again, showing that the current is only momentary. Let's say this again in different words. The battery keeps steadily forcing electrons through the circuit *ab* but the instrument in the circuit *cd* shows no current in that circuit except just at the instant when current starts to flow in the neighboring circuit *ab*.

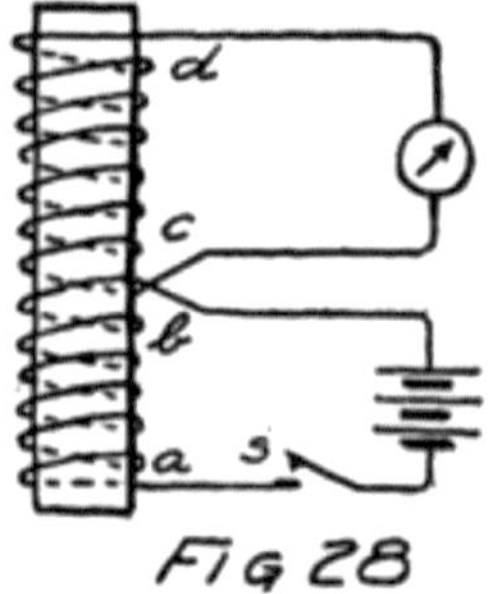

One thing this current-measuring instrument tells us is the direction of the electron stream through itself. It shows that the momentary stream of electrons goes through the coil from *d* to *c*, that is in the opposite direction to the stream in the part *ab*.

Now prepare to do a little close thinking. Read over carefully all I have told you about this experiment. You see that the moment the battery starts a stream of electrons from *a* towards *b*, something causes a momentary, that is a temporary, movement of electrons from *d* to *c*. We say that starting a 83 stream of electrons from *a* to *b* sets up or "induces" a stream of electrons from *d* to *c*.

What will happen then if we connect the battery between *a* and *d* as in Fig. 29? Electrons will start streaming away from *a* towards *b*, that is towards *d*. But that means there will be a momentary stream from *d* towards *c*, that is towards *a*. Our stream from the battery causes this oppositely directed stream. In the usual words we say it "induces" in the coil an opposing stream of electrons. This opposing stream doesn't last long, as we saw, but while it does last it hinders the stream which the battery is trying to establish.

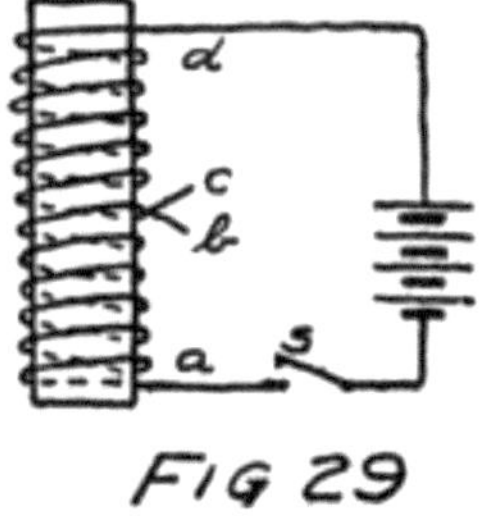

FIG 29

The stream of electrons which the battery causes will at first meet an opposition so it takes a little time before the battery can get the full-sized stream of electrons flowing steadily. In other words a current in a coil builds up slowly, because while it is building up it induces an effect which opposes somewhat its own building up.

Did you ever see a small boy start off somewhere, perhaps where he shouldn't be going, and find his conscience starting to trouble him at once. For a time he goes a little slowly but in a moment or two his conscience stops opposing him and he goes on steadily at his full pace. When he started he stirred up his conscience and that opposed him. Nobody else was hindering his going. It was all brought about by his own actions. The opposition which he 84met was "self-induced." He was hindered at first by a self-induced effect of his own conscience. If he was a stream of electrons starting off to travel around the coil we would say that he was opposed by a self-induced e. m. f. And any path in which such an effect will be produced we say has "self-inductance." Usually we shorten this term and speak of "inductance."

There is another way of looking at it. We know habits are hard to form and equally hard to break. It's hard to get electrons going around a coil and the self-inductance of a circuit tells us how hard it is. The harder it is the more self-inductance we say that the coil or circuit has. Of course, we need a unit in which to measure self-inductance. The unit is called the "henry." But that is more self-inductance than we can stand in most radio circuits, so we find it convenient to measure in smaller units called "mil-henries" which are thousandths of a henry.

You ought to know what a henry [4] is, if we are to use the word, but it isn't necessary just now to spend much time on it. The opposition which one's self-induced conscience offers depends upon how rapidly one starts. It's volts which make electrons move and so the conscience which opposes them will be measured in volts. Therefore we say that a coil has one henry of inductance when an electron stream 85which is increasing one ampere's worth each second stirs up in the coil a conscientious objection of one volt. Don't try to remember this now; you can come back to it later.

There is one more effect of inductance which we must know before we can get very far with our radio. Suppose an electron stream is flowing through a coil because a battery is driving the electrons along. Now let the battery be removed or disconnected. You'd expect the electron stream to stop at once but it doesn't. It keeps on for a moment because the electrons have got the habit.

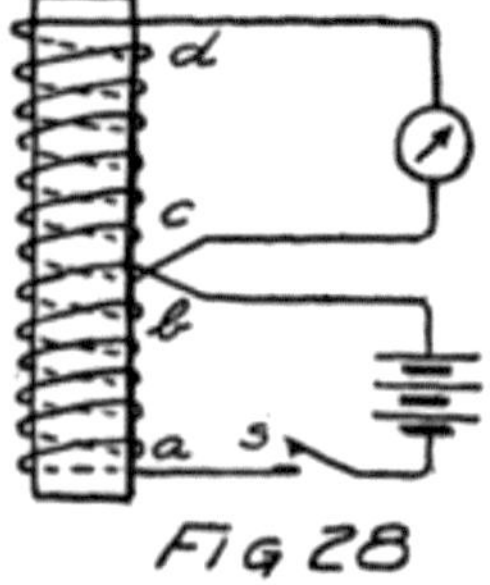

FIG 28

If you look again at Fig. 28 you will see what I mean. Suppose the switch is closed and a steady stream of electrons is flowing through the coil from *a* to *b*. There will be no current in the other part of the coil. Now open the switch. There will be a motion of the needle of the current-measuring instrument, showing a momentary current. The direction of this motion, however, shows that the momentary stream of electrons goes through the coil from *c* to *d*.

Do you see what this means? The moment the battery is disconnected there is nothing driving the electrons in the part *ab* and they slow down. Immediately, and just for an instant, a stream of electrons starts off in the part *cd* in the same direction as if the battery was driving them along.

86Now look again at Fig. 29. If the battery is suddenly disconnected there is a momentary rush of electrons in the same direction as the battery was driving them. Just as the self-inductance of a coil opposes the starting of a stream of electrons, so it opposes the stopping of a stream which is already going.

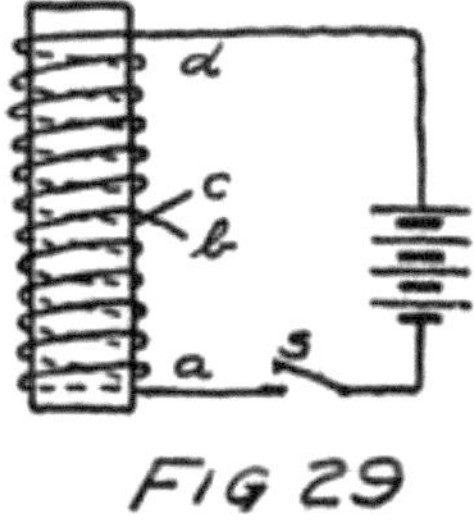

FIG 29

So far we haven't said much about making an audion produce alternating e. m. f.'s and thus making it useful for radio-telephony. Before radio was possible all these things that I have just told you, and some more too, had to be known. It took hundreds of good scientists years of patient study and experiment to find out those ideas about electricity which have made possible radio-telephony.

Two of these ideas are absolutely necessary for the student of radio-communication. First: A condenser is a gap in a circuit where there are waiting-rooms for the electrons. Second: Electrons form habits. It's hard to get them going through a coil of wire, harder than through a straight wire, but after they are going they don't like to stop. They like it much less if they are going through a coil instead of a straight wire.

In my next letter I'll tell you what happens when we have a coil and a condenser together in a circuit.

[4]

The "henry" has nothing to do with a well-known automobile. It was named after Joseph Henry, a professor years ago at Princeton University.

A "C-W" TRANSMITTER

DEAR SON:

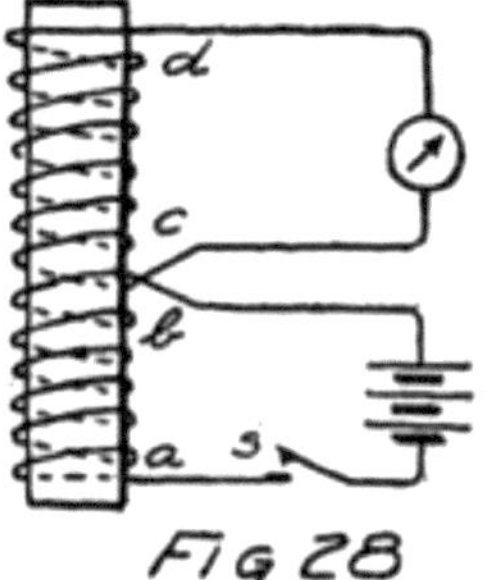

Let's look again at the coils of Fig. 28 which we studied in the last letter. I have reproduced them here so you won't have to turn back. When electrons start from *a* towards *b* there is a momentary stream of electrons from *d* towards *c*. If the electron stream through *ab* were started in the opposite direction, that is from *b* to *a* the induced stream in the coil *cd* would be from *c* towards *d*.

It all reminds me of two boys with a hedge or fence between them as in Fig. 30. One boy is after the other. Suppose you were being chased; you know what you'd do. If your pursuer started off 88with a rush towards one end of the hedge you'd "beat it" towards the other. But if he started slowly and cautiously you would start slow-ly too. You always go in the opposite direction, dodging back and forth along the paths which you are wearing in the grass on oppo-site sides of the hedge. If he starts to the right and then slows up and starts back, you will start to your right, slow up, and start back. Suppose he starts at the center of the hedge. First he dodges to the right, and then back through the center as far to the left, then back again and so on. You follow his every change.

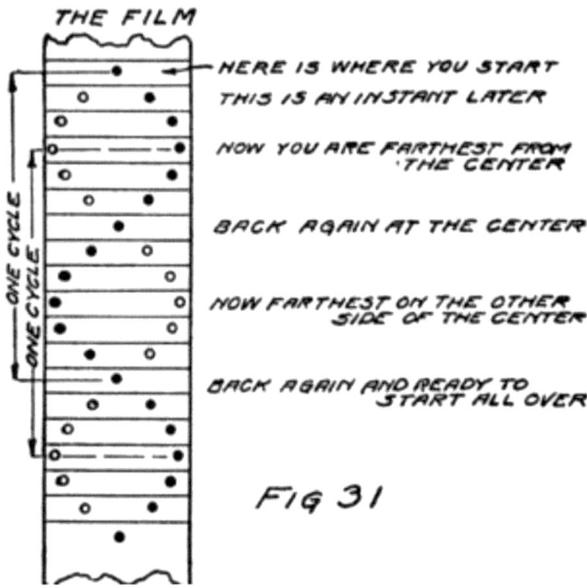

I am going to make a picture of what you two do. Let's start with the other fellow. He dodges or alternates back and forth. Some per-sons would say he "oscillates" back and forth in the same path. As 89he does so he induces you to move. I am on your side of the hedge with a moving-picture camera. My camera catches both of you. Fig. 31 shows the way the film would look if it caught only your heads. The white circle represents the tow-head on my side of the hedge and the black circle, young Brown who lives next door. Of course, the camera only catches you each time the shutter opens but it is easy to draw a complete picture of what takes place as time goes on. See Fig. 32.

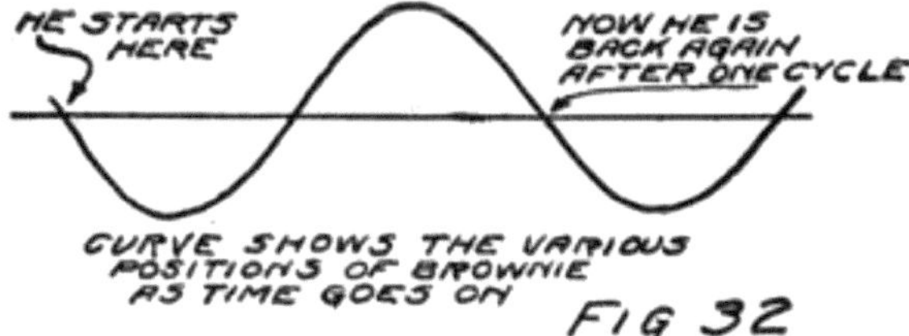

Now suppose you are an electron in coil *cd* of Fig. 33 and "Brownie" is one in coil *ab*. Your motions are induced by his. What's true of you two is true of all the other electrons. I have separated the coils a little in this sketch so that you can think of a hedge between. I don't know how one electron can affect another on the opposite side of this hedge but it can. And I don't know anything really about the hedge, which is generally called "the ether." The hedge isn't air. The effect would be the same if the coils were in a vacuum. The "ether" is just a name for whatever is left in the space about us when we have taken out everything 90 which we can see or feel–every molecule, every proton and every electron.

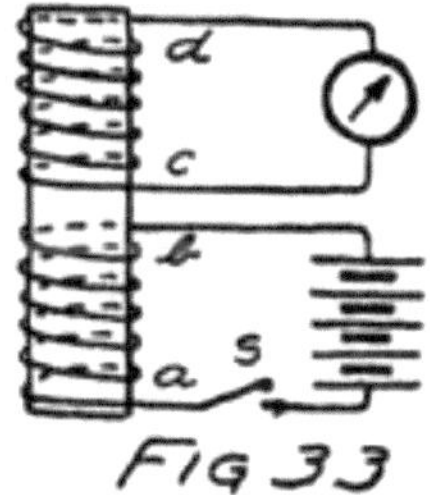

FIG 33

Why and how electrons can affect one another when they are widely separated is one of the great mysteries of science. We don't know any more about it than about why there are electrons. Let's accept it as a fundamental fact which we can't as yet explain.

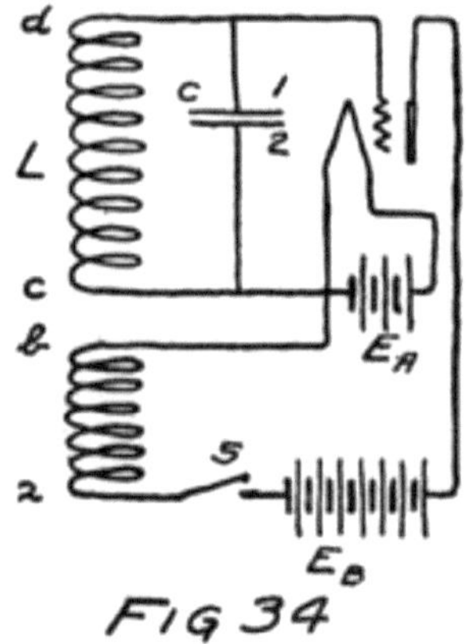

And now we can see how to make an audion produce an alternating current or as we sometimes say "make an audion oscillator." We shall set up an audion with its A-battery as in Fig. 34. Between the grid and the filament we put a coil and a condenser. Notice that they are in parallel, as we say. In the plate-filament circuit we connect the B-battery and a switch, S, and another coil. This coil in the plate circuit of the audion we place close to the other coil so that the two coils are just like the coils *ab* and *cd* of which I have been telling you. The moment any current flows in coil *ab* there will be a current flow in the coil *cd*. (An induced electron stream.) Of course, as long as the switch in the B-battery is open no current can flow.

The moment the switch S is closed the B-battery makes the plate positive with respect to the filament and there is a sudden surge of electrons round the 91 plate circuit and through the coil from *a* to *b*. You know what that does to the coil *cd*. It induces an electron stream from *d* towards *c*. Where do these electrons come from? Why, from the grid and the plate 1 of the condenser. Where do they go? Most of them go to the waiting-room offered by plate 2 of the condenser and some, of course, to the filament. What is the result? The grid becomes positive and the filament negative.

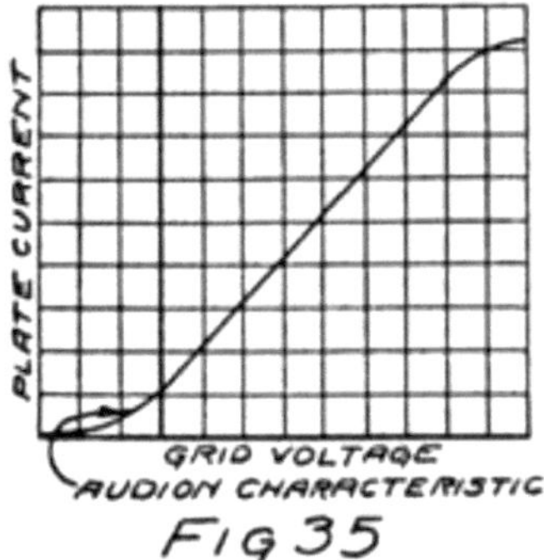

This is the crucial moment in our study. Can you tell me what is going to happen to the stream of electrons in the plate circuit? Remember that just at the instant when we closed the switch the grid was neither positive nor negative. We were at the point of zero volts on the audion characteristic of Fig. 35. When we close the switch the current in the plate circuit starts to jump from zero mil-amperes to the number of mil-amperes which represents the point where Zero Volt St. crosses Audion Characteristic. But this jump in plate current makes the grid positive as we have just seen. So the grid will help the plate call electrons and that will make the current in the plate circuit still larger, that is, result in a larger stream of electrons from *a* to *b*.

This increase in current will be matched by an increased effect in the coil *cd*, for you remember 92how you and "Brownie" behaved. And that will pull more electrons away from plate 1 of the condenser and send them to the waiting-room of 2. All this makes the grid more positive and so makes it call all the more effectively to help the plate move electrons.

Pl. V.–Variometer (top) and Variable Condenser (bottom) of the General Radio Company.

Voltmeter and Ammeter of the Weston Instrument Company.

We "started something" that time. It's going on all by itself. The grid is getting more positive, the plate current is getting bigger, and so the grid is getting more positive and the plate current still bigger. Is it ever going to stop? Yes. Look at the audion characteristic. There comes a time when making the grid a little more positive won't have any effect on the plate-circuit current. So the plate current stops increasing.

There is nothing now to keep pulling electrons away from plate 1 and crowding them into waiting-room 2. Why shouldn't the electrons in this waiting-room go home to that of plate 1? There is now no reason and so they start off with a rush.

Of course, some of them came from the grid and as fast as electrons get back to the grid it becomes less and less positive. As the grid becomes less and less positive it becomes less and less helpful to the plate.

If the grid doesn't help, the plate alone can't keep up this stream of electrons. All the plate can do by itself is to maintain the current represented by the intersection of zero volts and the audion characteristic. The result is that the current in the plate circuit, that is, of course, the current in coil *ab*, 93 becomes gradually less. About the time all the electrons, which had left the grid and plate 1 of the condenser, have got home the plate current is back to the value corresponding to $E_C = 0$.

The plate current first increases and then decreases, but it doesn't stop decreasing when it gets back to zero-grid value. And the reason is all due to the habit forming tendencies of electrons in coils. To see how this comes about, let's tell the whole story over again. In other words let's make a review and so get a sort of flying start.

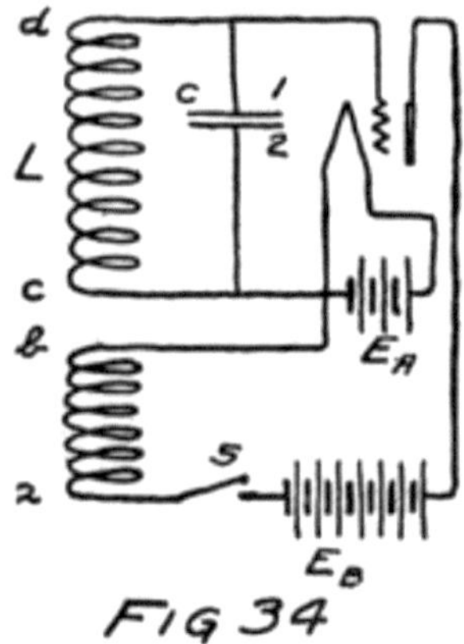

When we close the battery switch, *S* in Fig. 34, we allow a current to flow in the plate circuit. This current induces a current in the coil *cd* and charges the condenser which is across it, making plate 1 positive and plate 2 negative. A positive grid helps the plate so that the current in the plate circuit builds up to the greatest possible value as shown by the audion characteristic. That's the end of the increase in current. Now the condenser discharges, sending electrons through the coil *cd* and making the grid less positive until finally it is at zero potential, that is neither positive nor negative.

While the condenser is discharging the electrons in the coil *cd* get a habit of flowing from *c* toward *d*, that is from plate 2 to plate 1. If it wasn't for this 94 habit the electron stream in *cd* would stop as soon as the grid had reduced to zero voltage. Because of the habit, however, a lot of electrons that ought to stay on plate 2 get hurried along and land on plate 1. It is a little like the old game of "crack the whip." Some electrons get the habit and can't stop quickly enough so they go tumbling into waiting-room 1 and make it negative.

That means that the condenser not only discharges but starts to get charged in the other direction with plate 1 negative and plate 2 positive. The grid feels the effect of all this, because it gets extra electrons if plate 1 gets them. In fact the voltage effective between grid and filament is always the voltage between the plates of the condenser.

94

The audion characteristic tells us what is the result. As the grid becomes negative it opposes the plate, shooing electrons back towards the filament and reducing the plate current still further. But you have already seen in my previous letter what happens when we reduce the current in coil *ab*. There is then induced in coil *cd* an electron stream from *c* to *d*. This induced current is in just the right direction to send more electrons into waiting-room 1 and so to make the grid still more negative. And the more negative the grid gets the smaller becomes the plate current until finally the plate current is reduced to zero. Look at the audion characteristic again and see that making the grid sufficiently negative entirely stops the plate current.

When the plate current stops, the condenser in 95the grid circuit is charged, with plate 1 negative and 2 positive. It was the plate current which was the main cause of this change for it induced the charging current in coil *cd*. So, when the plate current becomes zero there is nothing to prevent the condenser from discharging.

Its discharge makes the grid less and less negative until it is zero volts and there we are–back practically where we started. The plate current is increasing and the grid is getting positive, and we're off on another "cycle" as we say. During a cycle the plate current increases to a maximum, decreases to zero, and then increases again to its initial value.

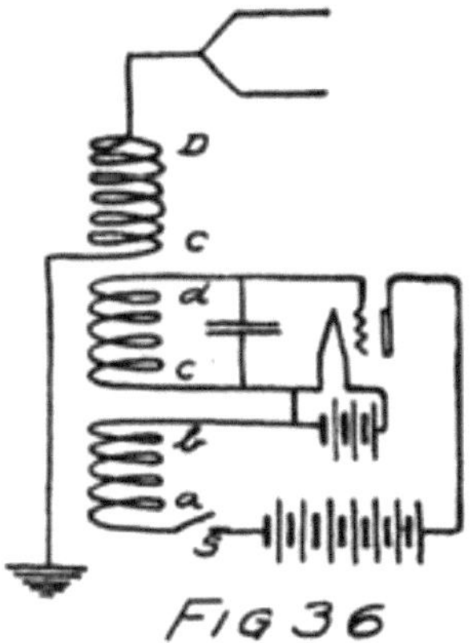

FIG 36

This letter has a longer continuous train of thought than I usually ask you to follow. But before I stop I want to give you some idea of what good this is in radio.

What about the current which flows in coil *cd*? It's an alternating current, isn't it? First the electrons stream from *d* towards *c*, and then back again from *c* towards *d*.

Suppose we set up another coil like *CD* in Fig. 36. It would have an alternating current induced in it. If this coil was connected to an antenna there would be radio waves sent out. The switch *S* could be used for a key and kept closed longer or shorter intervals 96 depending upon whether dashes or dots were being set. I'll tell you more about this later, but in this diagram are the makings of a "C-W Transmitter," that is a "continuous wave transmitter" for radio-telegraphy.

It would be worth while to go over this letter again using a pencil and tracing in the various circuits the electron streams which I have described.

97 LETTER 12
INDUCTANCE AND CAPACITY

DEAR SIR:

In the last letter I didn't stop to draw you a picture of the action of the audion oscillator which I described. I am going to do it now and you are to imagine me as using two pencils and drawing simultaneously two curves. One curve shows what happens to the current in the plate circuit. The other shows how the voltage of the grid changes. Both curves start from the instant when the switch is closed; and the two taken together show just what happens in the tube from instant to instant.

Fig. 37 shows the two curves. You will notice how I have drawn them beside and below the audion characteristic. The grid voltage and the plate current are related, as I have told you, and the audion characteristic is just a convenient way of showing the relationship. If we know the current in the plate circuit we can find the voltage of the grid and vice versa.

As time goes on, the plate current grows to its maximum and decreases to zero and then goes on climbing up and down between these two extremes. The grid voltage meanwhile is varying alternately, having its maximum positive value when the plate current is a maximum and its maximum negative 98value when the plate current is zero. Look at the two curves and see this for yourself.

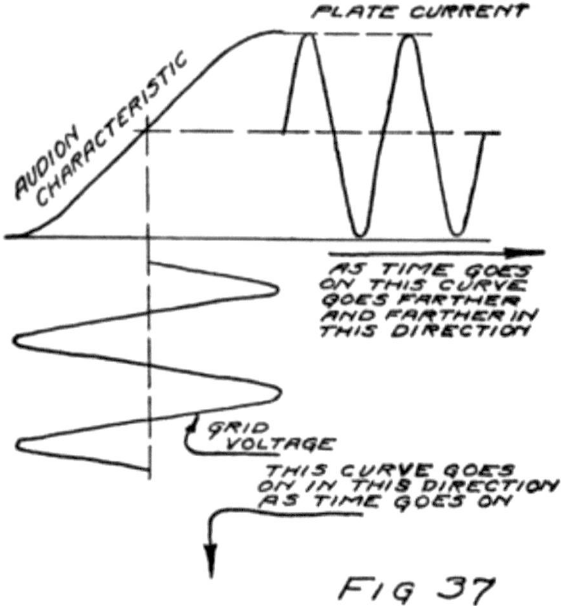

FIG 37

Now I want to tell you something about how fast these oscillations occur. We start by learning two words. One is "cycle" with which you are already partly familiar and the other is "frequency." Take cycle first. Starting from zero the current increases to a maximum, decreases to zero, and is ready again for the same series of changes. We say the current has passed through "a cycle of values." It doesn't make any difference where we start from. If we follow the current through all its different values until we are back at the same value as we started with and ready to start all over, then we have followed through a cycle of values.

99Once you get the idea of a cycle, and the markings on the curves in Fig. 31 will help you to understand, then the other idea is easy. By "frequency" we mean the number of cycles each second. The electric current which we use in lighting our house goes through sixty cycles a second. That means the current reverses its direction 120 times a second.

In radio we use alternating currents which have very high frequencies. In ship sets the frequency is either 500,000 or 1,000,000

cycles per second. Amateur transmitting sets usually have oscillators which run at well over a million cycles per second. The longer
range stations use lower frequencies.

You'll find, however, that the newspaper announcements of the
various broadcast stations do not tell the frequency but instead tell
the "wave length." I am not going to stop now to explain what that
means but I am going to give you a simple rule. Divide 300,000,000
by the "wave length" and you'll have the frequency. For example,
ships are supposed to use wave lengths of 300 meters or 600 meters.
Dividing three hundred million by three hundred gives one million
and that is one of the frequencies which I told you were used by
ship sets. Dividing by six hundred gives 500,000 or just half the
frequency. You can remember that sets transmitting with long
waves have low frequencies, but sets with short waves have high
frequencies. The frequency and the wave length don't change in the
same way. They change in opposite ways or inversely, 100as we
say. The higher the frequency the shorter the wave length.

I'll tell you about wave lengths later. First let's see how to control
the frequency of an audion oscillator like that of Fig. 38.

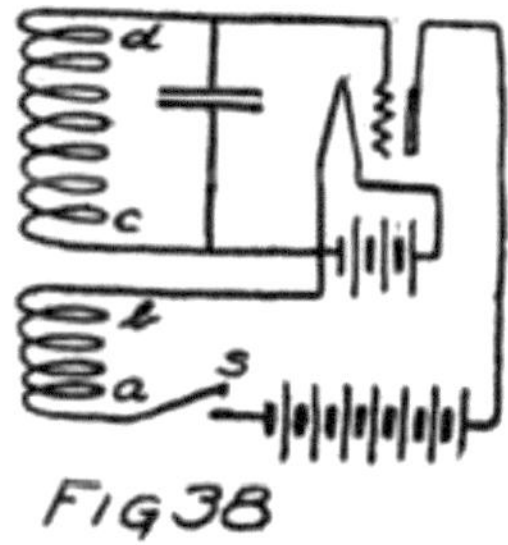

FIG 38

It takes time to get a full-sized stream going through a coil because of the inductance of the coil. That you have learned. And also
it takes time for such a current to stop completely. Therefore, if we
make the inductance of the coil small, keeping the condenser the
same, we shall make the time required for the current to start and
stop smaller. That will mean a higher frequency for there will be
more oscillations each second. One rule, then, for increasing the
frequency of an audion oscillator is to decrease the inductance.

Later in this letter I shall tell you how to increase or decrease the inductance of a coil. Before I do so, however, I want to call your attention to the other way in which we can change the frequency of an audion oscillator.

Let's see how the frequency will depend upon the capacity of the condenser. If a condenser has a large capacity it means that it can accommodate in its waiting-room a large number of electrons before the e. m. f. of the condenser becomes large enough to stop the stream of electrons which is charging the condenser. If the condenser in the grid circuit of Fig. 38 is of large capacity it means 101that it must receive in its upper waiting-room a large number of electrons before the grid will be negative enough to make the plate current zero. Therefore, the charging current will have to flow a long time to store up the necessary number of electrons.

You will get the same idea, of course, if you think about the electrons in the lower room. The current in the plate circuit will not stop increasing until the voltage of the grid has become positive enough to make the plate current a maximum. It can't do that until enough electrons have left the upper room and been stored away in the lower. Therefore the charging current will have to flow for a long time if the capacity is large. We have, therefore, the other rule for increasing the frequency of an audion oscillator, that is, decrease the capacity.

These rules can be stated the other way around. To decrease the frequency we can either increase the capacity or increase the inductance or do both.

But what would happen if we should decrease the capacity and increase the inductance? Decreasing the capacity would make the frequency higher, but increasing the inductance would make it lower. What would be the net effect? That would depend upon how much we decreased the capacity and how much we increased the inductance. It would be possible to decrease the capacity and then if we increased the inductance just the right amount to have no change in the frequency. No matter how large or how small we make the capacity we can 102always make the inductance such that there isn't any change in frequency. I'll give you a rule for this, after I have told you some more things about capacities and inductances.

First as to inductances. A short straight wire has a very small inductance, indeed. The longer the wire the larger will be the inductance but unless the length is hundreds of feet there isn't much inductance anyway. A coiled wire is very different.

A coil of wire will have more inductance the more turns there are to it. That isn't the whole story but it's enough for the moment. Let's see why. The reason why a stream of electrons has an opposing conscience when they are started off in a coil of wire is because each electron affects every other electron which can move in a parallel path. Look again at the coils of Figs. 28 and 29 which we discussed in the tenth letter. Those sketches plainly bring out the fact that the electrons in part *cd* travel in paths which are parallel to those of the electrons in part *ab*.

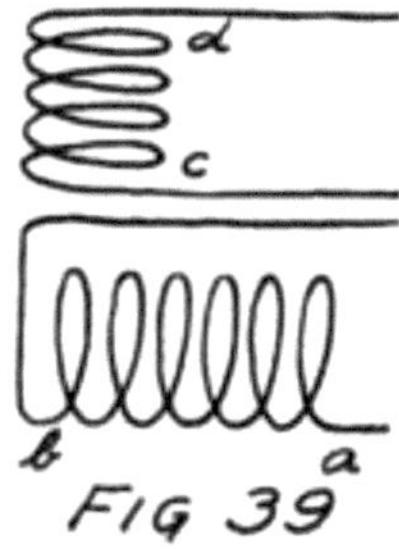

If we should turn these coils as in Fig. 39 so that all the paths in *cd* are at right angles to those in *ab* there wouldn't be any effect in *cd* when a current in *ab* started or stopped. Look at the circuit of the oscillating audion in Fig. 38. If we should turn these coils at right angles to each other we would stop the oscillation. Electrons only influence other electrons which are in parallel paths.

103When we want a large inductance we wind the coil so that there are many parallel paths. Then when the battery starts to drive an electron along, this electron affects all its fellows who are in parallel paths and tries to start them off in the opposite direction to that in which it is being driven. The battery, of course, starts to drive all the electrons, not only those nearest its negative terminal but those all along the wire. And every one of these electrons makes up for the fact that the battery is driving it along by urging all its fellows in the opposite direction.

It is not an exceptional state of affairs. Suppose a lot of boys are being driven out of a yard where they had no right to be playing. Suppose also that a boy can resist and lag back twice as much if some other boy urges him to do so. Make it easy and imagine three boys. The first boy lags back not only on his own account but because of the urging of the other boys. That makes him three times as hard to start as if the other boys didn't influence him. The same is true of the second boy and also of the third. The result is the unfortunate property owner has nine times as hard a job getting that gang started as if only one boy were to be dealt with. If there were two boys it would be four times as hard as for one boy. If there were four in the group it would be sixteen times, and if five it would be twenty-five times. The difficulty increases much more rapidly than the number of boys.

Now all we have to do to get the right idea of inductance 104is to think of each boy as standing for the electrons in one turn of the coil. If there are five turns there will be twenty-five times as much inductance, as for a single turn; and so on. You see that we can change the inductance of a coil very easily by changing the number of turns.

I'll tell you two things more about inductance because they will come in handy. The first is that the inductance will be larger if the turns are large circles. You can see that for yourself because if the circles were very small we would have practically a straight wire.

The other fact is this. If that property owner had been an electrical engineer and the boys had been electrons he would have fixed it so that while half of them said, "Aw, don't go; he can't put you off"; the other half would have said "Come on, let's get out." If he did that he would have a coil without any inductance, that is, he would have only the natural inertia of the electrons to deal with. We would say that he had made a coil with "pure resistance" or else that he had made a "non-inductive resistance."

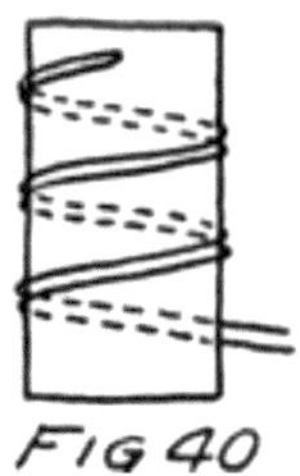

FIG 40

How would he do it? Easy enough after one learns how, but quite ingenious. Take the wire and fold it at the middle. Start with the middle and wind the coil with the doubled wire. Fig. 40 shows how the coil would look and you can see that part of the way the electrons are going around the coil in one direction and the rest of the way in the opposite 105direction. It is just as if the boys were paired off, a "goody-goody" and a "tough nut" together. They both shout at once opposite advice and neither has any effect.

I have told you all except one of the ways in which we can affect the inductance of a circuit. You know now all the methods which are important in radio. So let's consider how to make large or small capacities.

First I want to tell you how we measure the capacity of a condenser. We use units called "microfarads." You remember that an ampere means an electron stream at the rate of about six billion billion electrons a second. A millionth of an ampere would, therefore, be a stream at the rate of about six million million electrons a second– quite a sizable little stream for any one who wanted to count them as they went by. If a current of one millionth of an ampere should flow for just one second six million million electrons would pass along by every point in the path or circuit.

That is what would happen if there weren't any waiting-rooms in the circuit. If there was a condenser then that number of electrons would leave one waiting-room and would enter the other. Well, suppose that just as the last electron of this enormous number [5] entered its waiting-room we should know that the voltage of the condenser was just one volt. Then we would say that the condenser had a capacity of one microfarad. If it takes half that 106number to make the condenser oppose further changes in the contents of its waiting-rooms, with one volt's worth of opposition, that is, one volt of e. m. f., then the condenser has only half a microfarad of capacity. The number of microfarads of capacity (abbreviated mf.) is a measure of how many electrons we can get away from one plate and into the other before the voltage rises to one volt.

What must we do then to make a condenser with large capacity? Either of two things; either make the waiting-rooms large or put them close together.

If we make the plates of a condenser larger, keeping the separation between them the same, it means more space in the waiting-rooms and hence less crowding. You know that the more crowded the electrons become the more they push back against any other electron which some battery is trying to force into their waiting-room, that is the higher the e. m. f. of the condenser.

The other way to get a larger capacity is to bring the plates closer together, that is to shorten the gap. Look at it this way: The closer the plates are together the nearer home the electrons are. Their home is only just across a little gap; they can almost see the electronic games going on around the nuclei they left. They forget the long round-about journey they took to get to this new waiting-room and they crowd over to one side of this room to get just as close as they can to their old homes. That's why it's always easier, and takes less voltage, to get the same number of electrons moved from one plate to the 107other of a condenser which has only a small space between plates. It takes less voltage and that means that the condenser has a smaller e. m. f. for the same number of electrons. It also means that before the e. m. f. rises to one volt we can get more electrons moved around if the plates are close together. And that means larger capacity.

There is one thing to remember in all this: It doesn't make any difference how thick the plates are. It all depends upon how much surface they have and how close together they are. Most of the electrons in the plate which is being made negative are way over on the side toward their old homes, that is, toward the plate which is being made positive. And most of the homes, that is, atoms which have lost electrons, are on the side of the positive plate which is next to the gap. That's why I said the electrons could almost see their old homes.

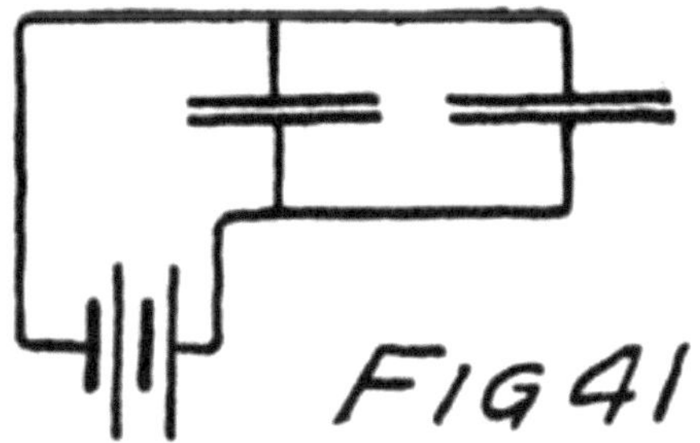

All this leads to two very simple rules for building condensers. If you have a condenser with too small a capacity and want one, say, twice as large, you can either use twice as large plates or bring the plates you already have twice as close together; that is, make the gap half as large. Generally, of course, the 108gap is pretty well fixed. For example, if we make a condenser by using two pieces of metal and separating them by a sheet of mica we don't want the job of splitting the mica. So we increase the size of the plates. We can do that either by using larger plates or other plates and connecting it as in Fig. 41 so that the total waiting-room space for electrons is increased.

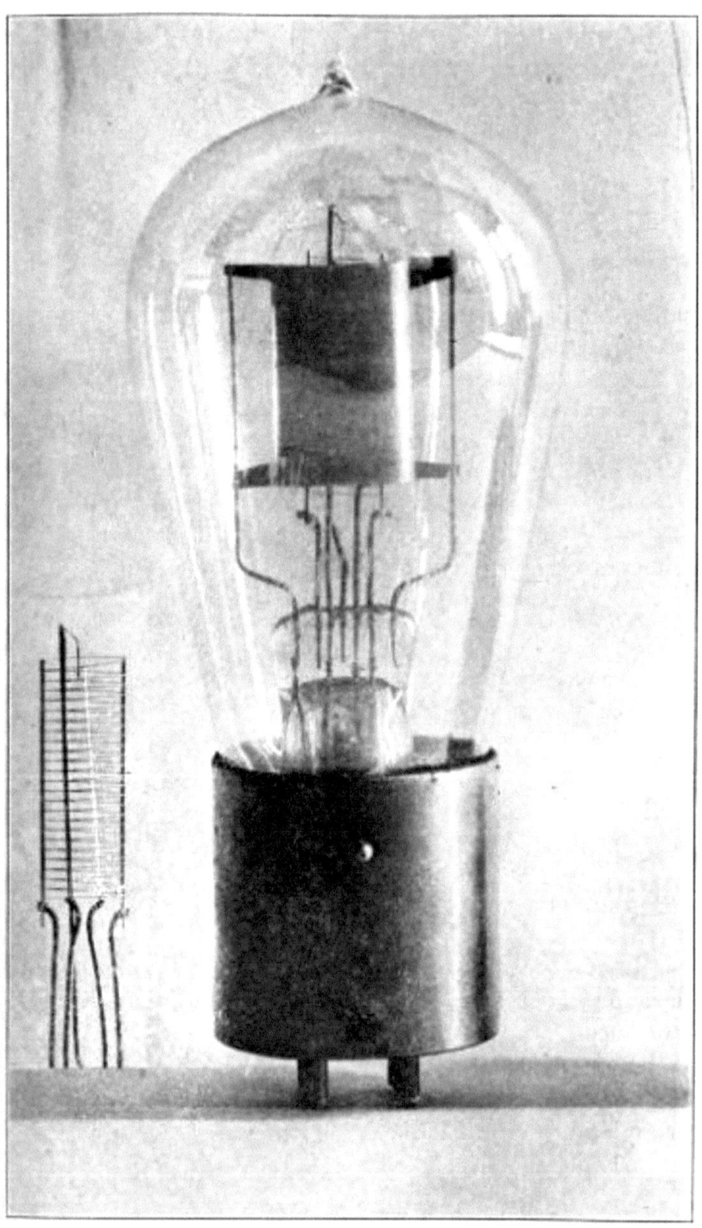

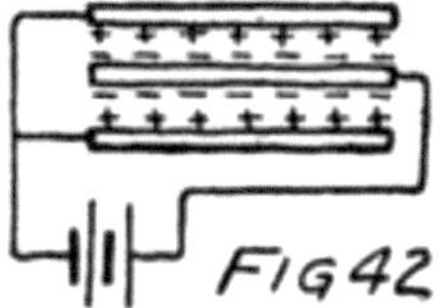

109If you have got these ideas you can understand how we use both sides of the same plate in some types of condensers. Look at Fig. 42. There are two plates connected together and a third between them. Suppose electrons are pulled from the outside plates and crowded into the middle plate. Some of them go on one side and some on the other, as I have shown. The negative signs indicate electrons and the plus signs their old homes. If we use more plates as in Fig. 43 we have a larger capacity.

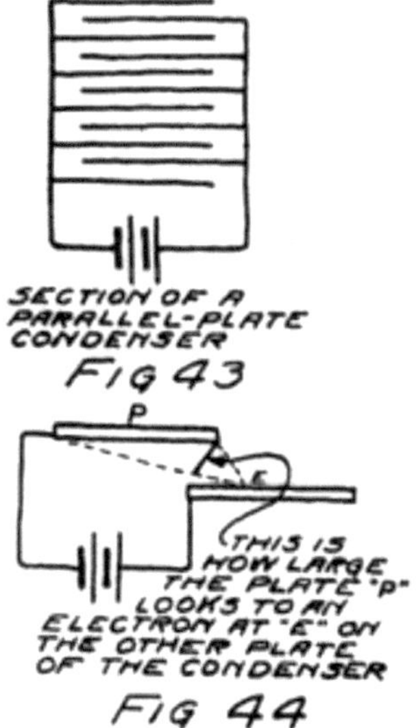

What if we have two plates which are not directly opposite one another, like those of Fig. 44? What does the capacity depend upon? Imagine yourself an electron on the negative plate. Look off toward the positive plate and see how big it seems to you. The bigger it looks the more capacity the condenser has. When the plates are right opposite one another the positive plate looms up pretty large. But if they slide apart you don't see so much of it; and if it is off to one side about all you 110see is the edge. If you can't see lots of atoms which have lost electrons and so would make good homes for you, there is no use of your staying around on that side of the plate; you might just as well be trying to go back home the long way which you originally came.

That's why in a variable plate condenser there is very little capacity when no parts of the plates are opposite each other, and there is the greatest capacity when they are exactly opposite one another.

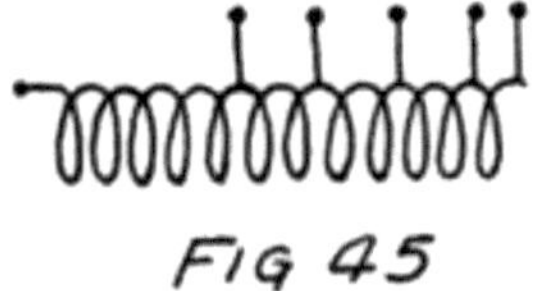

FIG 45

While we are at it we might just as well clean up this whole business of variable capacities and inductances by considering two ways in which to make a variable inductance. Fig. 45 shows the simplest way but it has some disadvantages which I won't try now to explain. We make a long coil and then take off taps. We can make connections between one end of the coil and any of the taps. The more turns there are included in the part of the coil which we are using the greater is the inductance. If we want to do a real job we can bring each of these taps to a little stud and arrange a sliding or rotating contact with them. Then we have an inductance the value of which we can vary "step-by-step" in a convenient manner.

Another way to make a variable inductance is to make what is called a "variometer." I dislike the name because it doesn't "meter" anything. If properly calibrated it would of course "meter" inductance, 111but then it should be called an "inducto-meter."

Do you remember the gang of boys that fellow had to drive off his property? What if there had been two different gangs playing there? How much trouble he has depends upon whether there is anything in common between the gangs. Suppose they are playing in different parts of his property and so act just as if the other crowd wasn't also trespassing. He could just add the trouble of starting one gang to the trouble of starting the other.

108

It would be very different if the gangs have anything in common. Then one would encourage the other much as the various boys of the same gang encourage each other. He would have a lot more trouble. And this extra trouble would be because of the relations between gangs, that is, because of their "mutual inductance."

On the other hand suppose the gangs came from different parts of the town and disliked each other. He wouldn't have nearly the trouble. Each gang would be yelling at the other as they went along: "You'd better beat it. He knows all right, all right, who broke that bush down by the gate. Just wait till he catches you." They'd get out a little easier, each in the hope the other crowd would catch it from the owner. There's a case where their mutual relations, their mutual inductance, makes the job easier.

That's true of coils with inductance. Suppose you wind two inductance coils and connect them in series. If they are at right angles to each other as in Fig. 11246a they have no effect on each other. There is no mutual inductance. But if they are parallel and wound the same way like the coils of Fig. 46b they will act like a single coil of greater inductance. If the coils are parallel but wound in opposite directions as in Fig. 46c they will have less inductance because of their mutual inductance. You can check these statements for yourself if you'll refer back to Letter 10 and see what happens in the same way as I told you in discussing Fig. 28.

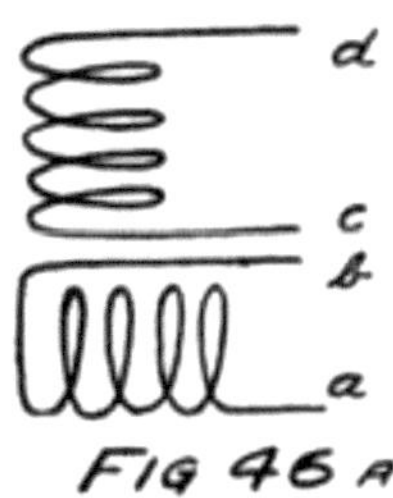

FIG 46 A

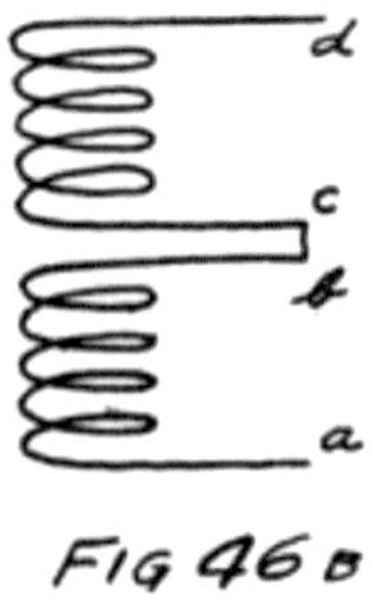

FIG 46 b

If the coils are neither parallel nor at right angles there will be some mutual inductance but not as much as if they were parallel. By turning the coils we can get all the variations in mutual relations from the case of Fig. 46b to that of Fig. 46c. That's what we arrange to do in a variable inductance of the variometer type.

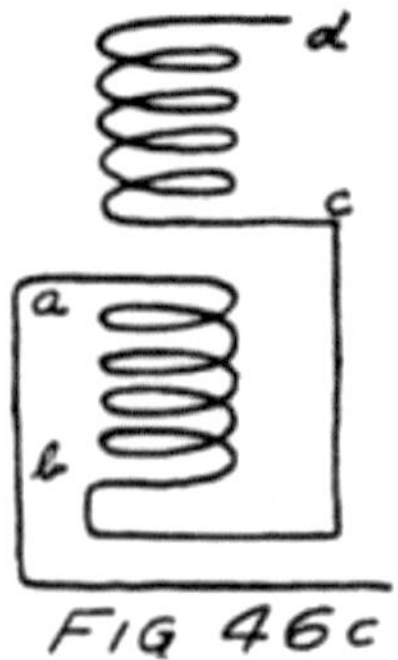

FIG 46c

There is another way of varying the mutual inductance. We can make one coil slide inside another. If it is way inside, the total inductance which the two coils offer is either larger than the sum of what they can offer separately or less, depending upon whether the windings are in 113the same direction or opposite. As we pull the coil out the mutual effect becomes less and finally when it is well outside the mutual inductance is very small.

Now we have several methods of varying capacity and inductance and therefore we are ready to vary the frequency of our audion oscillator; that is, "tune" it, as we say. In my next letter I shall show you why we tune.

Now for the rule which I promised. The frequency to which a circuit is tuned depends upon the product of the number of milhenries in the coil and the number of microfarads in the condenser. Change the coil and the condenser as much as you want but keep this product the same and the frequency will be the same.

[5]

More accurately the number is 6,286,000,000,000.

114 LETTER 13
TUNING

DEAR RADIO ENTHUSIAST:

I want to tell you about receiving sets and their tuning. In the last letter I told you what determines the frequency of oscillation of an audion oscillator. It was the condenser and inductance which you studied in connection with Fig. 36. That's what determines the frequency and also what makes the oscillations. All the tube does is to keep them going. Let's see why this is so.

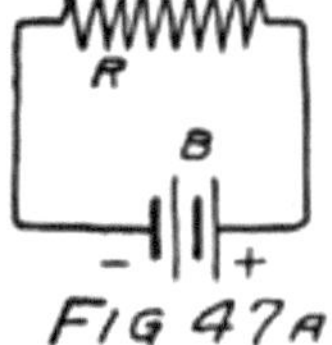

Start first, as in Fig. 47a, with a very simple circuit of a battery and a non-inductive resistance, that is, a wire wound like that of Fig. 40 in the previous letter, so that it has no inductance. The battery must do work forcing electrons through that wire. It has the ability, or the energy as we say.

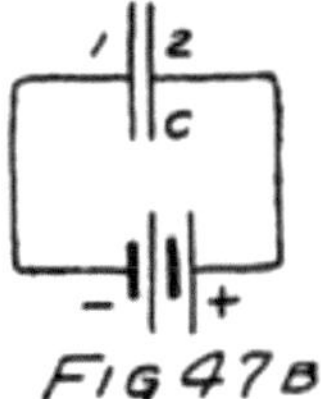

Now connect a condenser to the battery as in Fig. 47b. The connecting wires are very short; and so practically all the work which the battery does is in storing electrons in the negative plate of the condenser and robbing the positive plate. The battery displaces a certain number of electrons in the waiting-rooms of the condenser. How many, depends upon how hard it 115can push and pull, that is on its e. m. f., and upon how much capacity the condenser has.

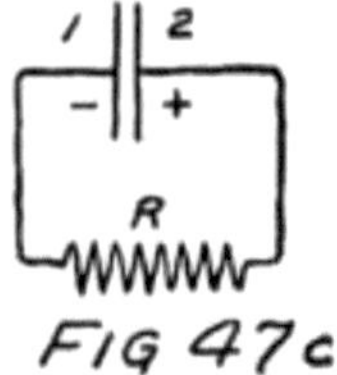

FIG 47c

Remove the battery and connect the charged condenser to the resistance as in Fig. 47c. The electrons rush home. They bump and jostle their way along, heating the wire as they go. They have a certain amount of energy or ability to do work because they are away from home and they use it all up, bouncing along on their way. When once they are home they have used up all the surplus energy which the battery gave them.

Try it again, but this time, as in Fig. 47d, connect the charged condenser to a coil which has inductance. The electrons don't get started as fast because of the inductance. But they keep going because the electrons in the wire form the habit. The result is that about the time enough electrons have got into plate 2 (which was positive), to satisfy all its lonely protons, the electrons in the wire are streaming along at a great rate. A lot of them keep going until they land on this plate and so make it negative.

FIG 47d

That's the same sort of thing that happens in the case of the inductance and condenser in the oscillating audion circuit except for one important fact. There is nothing to keep electrons going to the 2 plate except this habit. And there are plenty of stay-at-home electrons to stop them as they rush along. They bump and jostle, but some of them are stopped or else diverted so 116that they go bumping around without getting any nearer plate 2. Of course, they spend all their energy this way, getting every one all stirred up and heating the wire.

Some of the energy which the electrons had when they were on plate 1 is spent, therefore, and there aren't as many electrons getting to plate 2. When they turn around and start back, as you know they do, the same thing happens. The result is that each successive surge of electrons is smaller than the preceding. Their energy is being wasted in heating the wire. The stream of electrons gets smaller and smaller, and the voltage of the condenser gets smaller and smaller, until by-and-by there isn't any stream and the condenser is left uncharged. When that happens, we say the oscillations have "damped out."

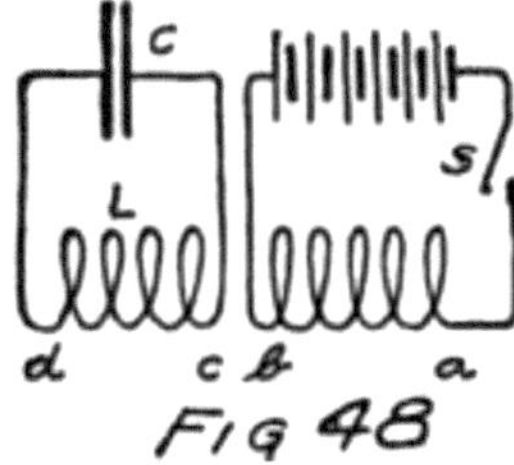

That's one way of starting oscillations which damp out–to start with a charged condenser and connect an inductance across it. There is another way which leads us to some important ideas. Look at Fig. 48. There is an inductance and a condenser. Near the coil is another coil which has a battery and a key in circuit with it. The coils are our old friends of Fig. 33 in Letter 10. Suppose we close the switch S. It starts a current through the coil *ab* which goes on steadily as soon as it really gets going. While it is starting, however, it induces an electron stream 117 in coil *cd*. There is only a momentary or transient current but it serves to charge the condenser and then events happen just as they did in the case where we charged the condenser with a battery.

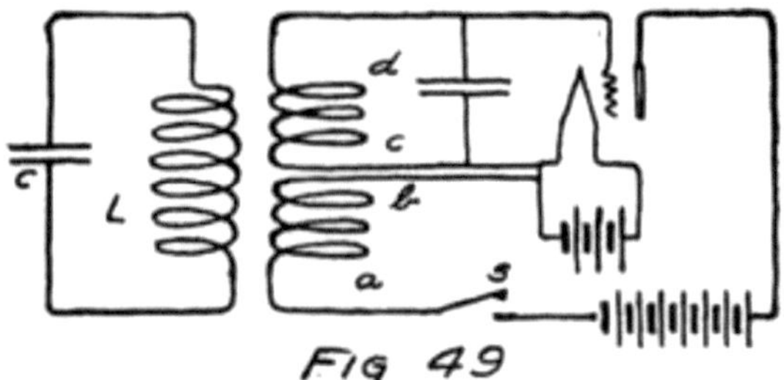

Now take away this coil *ab* with its battery and substitute the oscillator of Fig. 36. What's going to happen? We have two circuits in which oscillations can occur. See Fig. 49. One circuit is associated with an audion and some batteries which keep supplying it with energy so that its oscillations are continuous. The other circuit is near enough to the first to be influenced by what happens in that circuit. We say it is "coupled" to it, because whatever happens in the first circuit induces an effect in the second circuit.

Suppose first that in each circuit the inductance and capacity have such values as to produce oscillations of the same frequency. Then the moment we start the oscillator we have the same effect in both circuits. Let me draw the picture a little differently (Fig. 50) so that you can see this more easily. I have merely made the coil *ab* in two parts, one of which can affect *cd* in the oscillator and the other the coil *L* of the second circuit.

118But suppose that the two circuits do not have the same natural frequencies, that is the condenser and inductance in one circuit are so large that it just naturally takes more time for an oscillation in that circuit than in the other. It is like learning to dance. You know about how well you and your partner would get along if you had one frequency of oscillation and she had another. That's what happens in a case like this.

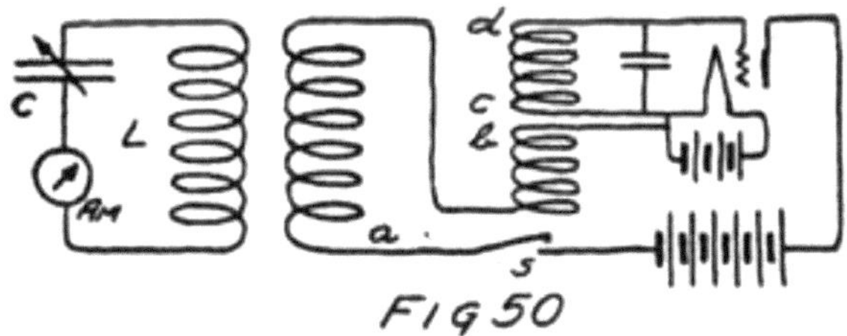

FIG 50

If circuit *L-C* takes longer for each oscillation than does circuit *ab* its electron stream is always working at cross purposes with the electron stream in *ab* which is trying to lead it. Its electrons start off from one condenser plate to the other and before they have much more than got started the stream in *ab* tries to call them back to go in the other direction. It is practically impossible under these conditions to get a stream of any size going in circuit *L-C*. It is equally hard if *L-C* has smaller capacity and inductance than *ab* so that it naturally oscillates faster.

I'll tell you exactly what it is like. Suppose you and your partner are trying to dance without any piano or other source of music. She has one tune running through her head and she dances to that, 119except as you drag her around the floor. You are trying to follow another tune. As a couple you have a difficult time going anywhere under these conditions. But it would be all right if you both had the same tune.

If we want the electron stream in coil *ab* to have a large guiding effect on the stream in coil *L-C* we must see that both circuits have the same tune, that is the same natural frequency of oscillation.

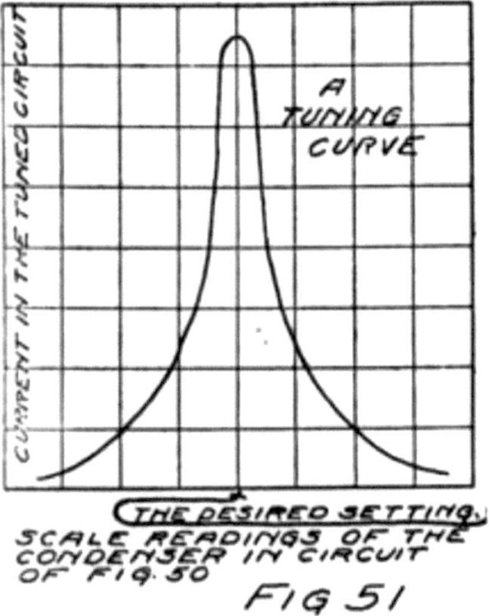

FIG 51

This can be shown very easily by a simple experiment. Suppose we set up our circuit *L-C* with an ammeter in it, so as to be able to tell how large an electron stream is oscillating in that circuit. Let us also make the condenser a variable one so that we can change the natural frequency or tune of the circuit. Now let's see what happens to the current as we vary this condenser, changing the capacity and thus changing the tune of the circuit. If we use a variable plate condenser it will have a scale on top graduated in degrees and we can note the reading of the ammeter for each position of the movable 120 plates. If we do, we find one position of these plates, that is one setting, corresponding to one value of capacity in the condenser, where the current in the circuit is a maximum. This is the setting of the condenser for which the circuit has the same tune or natural frequency as the circuit *cd*. Sometimes we say that the circuits are now in resonance. We also refer to the curve of values of current and condenser positions as a "tuning curve." Such a curve is shown in Fig. 51.

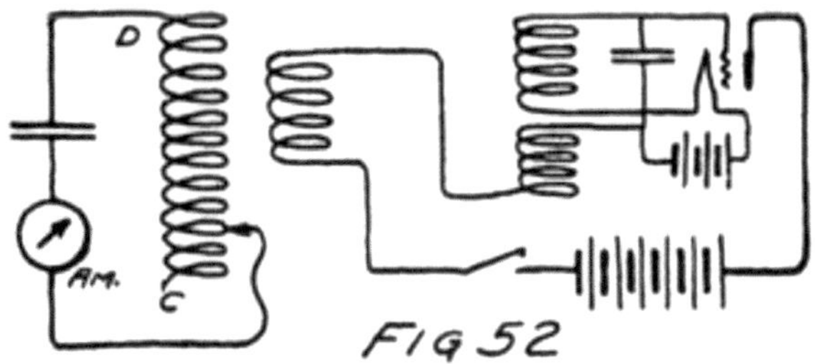

That's all there is to tuning–adjusting the capacity and inductance of a circuit until it has the same natural frequency as some other circuit with which we want it to work. We can either adjust the capacity as we just did, or we can adjust the inductance. In that case we use a variable inductance as in Fig. 52.

If we want to be able to tune to any of a large range of frequencies we usually have to take out or put into the circuit a whole lot of mil-henries at a time. When we do we get these mil-henries of induct-ance from a coil which we call a "loading coil." That's why your friends add a loading coil when they 121want to tune for the long wave-length stations, that is, those with a low frequency.

When our circuit L-C of Fig. 49 is tuned to the frequency of the oscillator we get in it a maximum current. There is a maximum stream of electrons, and hence a maximum number of them crowd-ed first into one and then into the other plate of the condenser. And so the condenser is charged to a maximum voltage, first in one di-rection and then in the other.

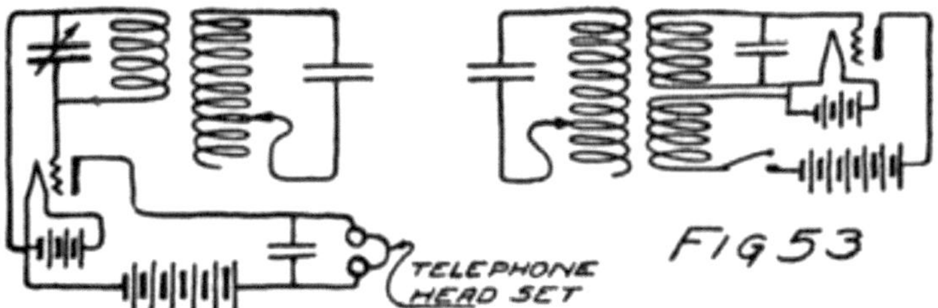

Now connect the circuit L-C to the grid of an audion. If the circuit is tuned we'll have the maximum possible voltage applied between grid and filament. In the plate circuit we'll get an increase and then a decrease of current. You know that will happen for I prepared you for this moment by the last page of my ninth letter. I'll tell you more about that current in the plate circuit in a later letter. I am connecting a telephone receiver in the plate circuit, and also a condenser, the latter for a reason to be explained later. The combination appears then as in Fig. 53. That figure shows a C-W transmitter and an audion detector. This is the sort of a detector 122 we would use for radio-telephony, but the transmitter is the sort we would use for radio-telegraphy. We shall make some changes in them later.

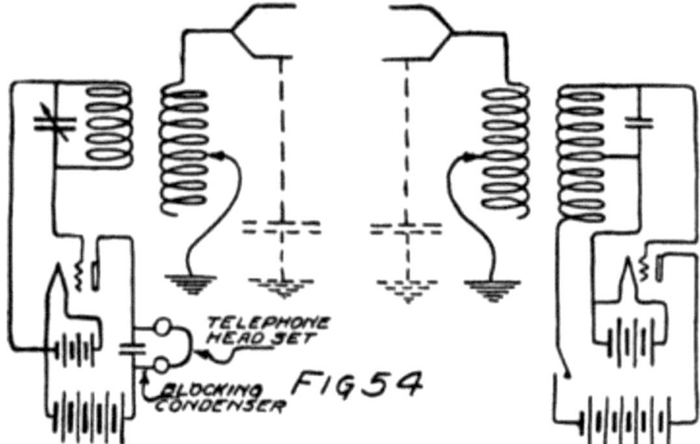

Whenever we start the oscillating current in the transmitter we get an effect in the detector circuit, of which I'll tell you more later. For the moment I am interested in showing you how the transmitter and the detector may be separated by miles and still there will be an effect in the detector circuit every time the key in the transmitter circuit is closed.

This is how we do it. At the sending station, that is, wherever we locate the transmitter, we make a condenser using the earth, or ground, as one plate. We do the same thing at the receiving station where the detector circuit is located. To these condensers we connect inductances and these inductances we couple to our transmitter and receiver as shown in Fig. 54. The upper plate of the condenser in each 123case is a few horizontal wires. The lower plate is the moist earth of the ground and we arrange to get in contact with that

120

in various ways. One of the simplest methods is to connect to the water pipes of the city water-system.

Now we have our radio transmitting-station and a station for receiving its signals. You remember we can make dots and dashes by the key or switch in the oscillator circuit. When we depress the key we start the oscillator going. That sets up oscillations in the circuit with the inductance and the capacity formed by the antenna. If we want a real-sized stream of electrons up and down this antenna lead (the vertical wire), we must tune that circuit. That is why I have shown a variable inductance in the circuit of the transmitting antenna.

What happens when these electrons surge back and forth between the horizontal wires and the ground, I don't know. I do know, however, that if we tune the antenna circuit at the receiving station there will be a small stream of electrons surging back and forth in that circuit.

Usually scientists explain what happens by saying that the transmitting station sends out waves in the ether and that these waves are received by the antenna system at the distant station. Wherever you put up a receiving station you will get the effect. It will be much smaller, however, the farther the two stations are apart.

I am not going to tell you anything about wave motion in the ether because I don't believe we know 124enough about the ether to try to explain, but I shall tell you what we mean by "wave length."

Somehow energy, the ability to do work, travels out from the sending antenna in all directions. Wherever you put up your receiving station you get more or less of this energy. Of course, energy is being sent out only while the key is depressed and the oscillator going. This energy travels just as fast as light, that is at the enormous speed of 186,000 miles a second. If you use meters instead of miles the speed is 300,000,000 meters a second.

Now, how far will the energy which is sent out from the antenna travel during the time it takes for one oscillation of the current in the antenna? Suppose the current is oscillating one million times a second. Then it takes one-millionth of a second for one oscillation. In that time the energy will have traveled away from the antenna one-millionth part of the distance it will travel in a whole second. That is one-millionth of 300 million meters or 300 meters.

The distance which energy will go in the time taken by one oscillation of the source of that energy is the wave length. In the case just given that distance is 300 meters. The wave length, then, of 300 meters corresponds to a frequency of one million. In fact if we divide 300 million meters by the frequency we get the wave length, and that's the same rule as I gave you in the last letter.

In further letters I'll tell you how the audion works as a detector and how we connect a telephone 125transmitter to the oscillator to make it send out energy with a speech significance instead of a mere dot and dash significance, or signal significance. We shall have to learn quite a little about the telephone itself and about the human voice.

126 LETTER 14
WHY AND HOW TO USE A DETECTOR

Dear Son:

In the last letter we got far enough to sketch, in Fig. 54, a radio transmitting station and a receiving station. We should never, however, use just this combination because the transmitting station is intended to send telegraph signals and the receiving station is best suited to receiving telephonic transmission. But let us see what happens.

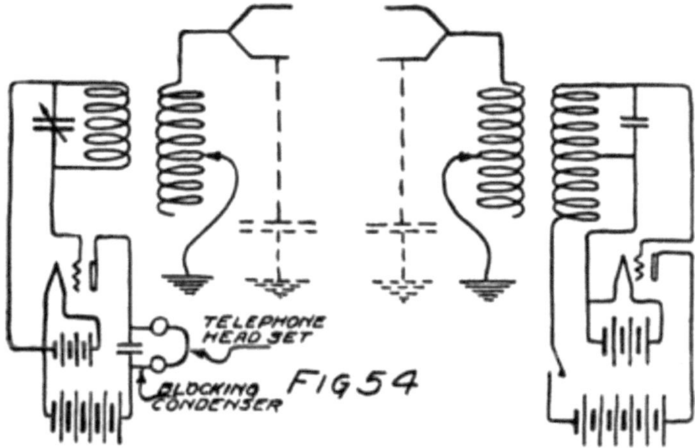

When the key in the plate circuit of the audion at the sending station is depressed an alternating current is started. This induces an alternating current in the neighboring antenna circuit. If this antenna circuit, which is formed by a coil and a condenser, is tuned to the frequency of oscillations which are 127being produced in the audion circuit then there is a maximum current induced in the antenna.

As soon as this starts the antenna starts to send out energy in all directions, or "radiate" energy as we say. How this energy, or ability to do work, gets across space we don't know. However it may be, it does get to the receiving station. It only takes a small fraction of a second before the antenna at the receiving station starts to receive energy, because energy travels at the rate of 186,000 miles a second.

The energy which is received does its work in making the electrons in that antenna oscillate back and forth. If the receiving antenna is tuned to the frequency which the sending station is producing, then the electrons in the receiving antenna oscillate back and forth most widely and there is a maximum current in this circuit.

The oscillations of the electrons in the receiving antenna induce similar oscillations in the tuned circuit which is coupled to it. This circuit also is tuned to the frequency which the distant oscillator is producing and so in it we have the maximum oscillation of the electrons. The condenser in that circuit charges and discharges alternately.

The grid of the receiving audion always has the same voltage as the condenser to which it is connected and so it becomes alternately positive and negative. This state of affairs starts almost as soon as the key at the sending station is depressed and continues as long as it is held down.

Now what happens inside the audion? As the 128grid becomes more and more positive the current in the plate circuit increases. When the grid no longer grows more positive but rather becomes less and less positive the current in the plate circuit decreases. As the grid becomes of zero voltage and then negative, that is as the grid "reverses its polarity," the plate current continues to decrease. When the grid stops growing more negative and starts to become less so, the plate current stops decreasing and starts to increase.

All this you know, for you have followed through such a cycle of changes before. You know also how we can use the audion characteristic to tell us what sort of changes take place in the plate current when the grid voltage changes. The plate current increases and decreases alternately, becoming greater and less than it would be if the grid were not interfering. These variations in its intensity take place very rapidly, that is with whatever high frequency the sending station operates. What happens to the plate current on the average?

The plate current, you remember, is a stream of electrons from the filament to the plate (on the inside of the tube), and from the plate back through the B-battery to the filament (on the outside of the tube). The grid alternately assists and opposes that stream. When it assists, the electrons in the plate circuit are moved at a faster rate. When the grid becomes negative and opposes the plate the stream of electrons is at a slower rate. The stream is always going in the same direction but it varies in its 129rate depending upon the changes in grid potential.

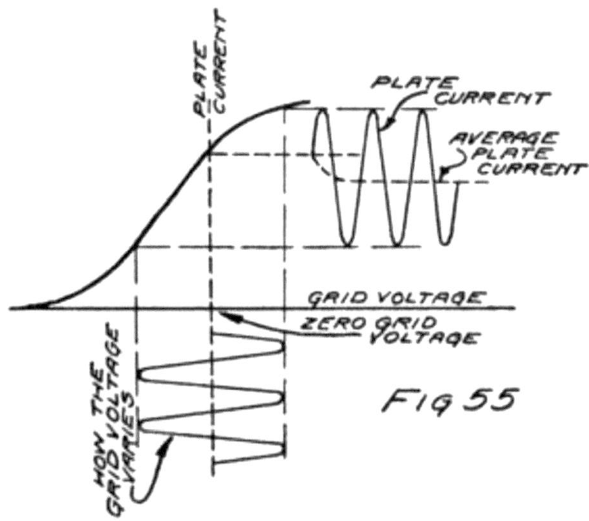

When the grid is positive, that is for half a cycle of the alternating grid-voltage, the stream is larger than it would be if the plate current depended only on the B-battery. For the other half of a cycle it is less. The question I am raising is this: Do more electrons move around the plate circuit if there is a signal coming in than when there is no incoming signal? To answer this we must look at the audion characteristic of our particular tube and this characteristic must have been taken with the same B-battery as we use when we try to receive the signals.

There are just three possible answers to this question. The first answer is: "No, there is a smaller number of electrons passing through the plate circuit each second if the grid is being affected by

an incoming signal." The second is: "The signal 130doesn't make
any difference in the total number of electrons which move each
second from filament to plate." And the third answer is: "Yes, there
is a greater total number each second."

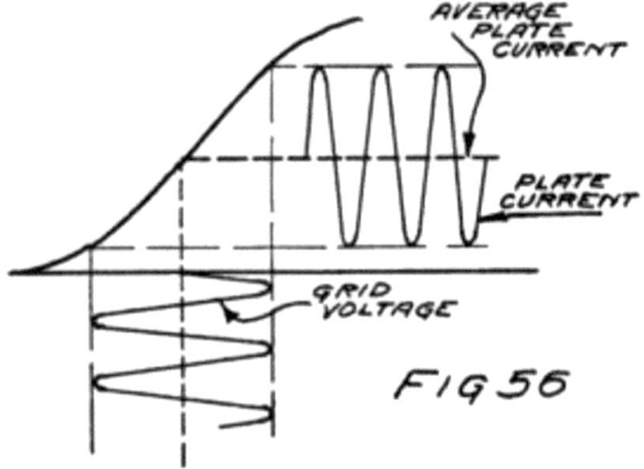

Any one of the three answers may be right. It all depends on the
characteristic of the tube as we are operating it, and that depends
not only upon the type and design of tube but also upon what voltages we are using in our batteries. Suppose the variations in the
voltage of the grid are as represented in Fig. 55, and that the characteristic of the tube is as shown in the same figure. Then obviously
the first answer is correct. You can see for yourself that when the
grid becomes positive the current in the plate circuit can't increase
much anyway. For the other half of the cycle, that is, while the grid
is negative, the current in the plate is very much decreased. The
decrease in one half-cycle is larger than the increase during the other half-cycle, so that on the average the current is less when the
131signal is coming in. The dotted line shows the average current.

Suppose that we take the same tube and use a B-battery of lower
voltage. The characteristic will have the same shape but there will
not be as much current unless the grid helps, so that the characteristic will be like that of Fig. 56. This characteristic crosses the axis of
zero volts at a smaller number of mil-amperes than does the other
because the B-batteries can't pull as hard as they did in the other
case.

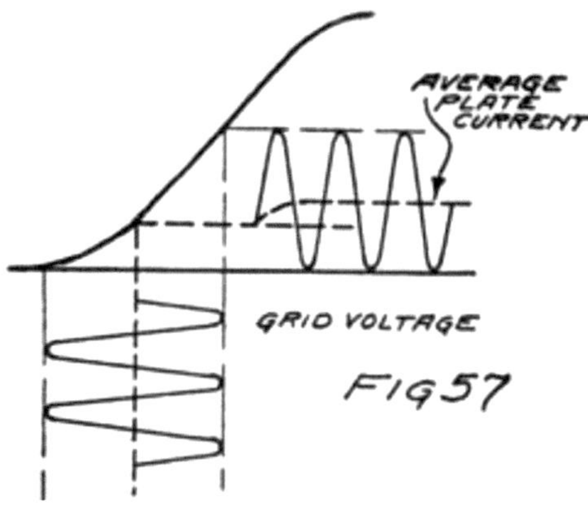

You can see the result. When the grid becomes positive it helps and increases the plate current. When it becomes negative it opposes and decreases the plate current. But the increase just balances the decrease, so that on the average the current is unchanged, as shown by the dotted line.

On the other hand, if we use a still smaller voltage of B-battery we get a characteristic which shows a still smaller current when the grid is at zero potential. 132For this case, as shown in Fig. 57, the plate current is larger on the average when there is an incoming signal.

If we want to know whether or not there is any incoming signal we will not use the tube in the second condition, that of Fig. 56, because it won't tell us anything. On the other hand why use the tube under the first conditions where we need a large plate battery? If we can get the same result, that is an indication when the other station is signalling, by using a small battery let's do it that way for batteries cost money. For that reason we shall confine ourselves to the study of what takes place under the conditions of Fig. 57.

We now know that when a signal is being sent by the distant station the current in the plate circuit of our audion at the receiving station is greater, on the average. We are ready to see what effect this has on the telephone receiver. And to do this requires a little study of how the telephone receiver works and why.

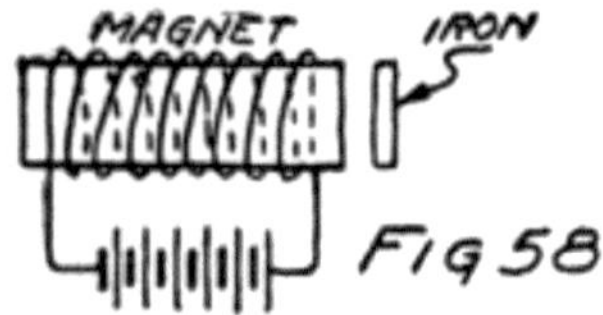

I shall not stop now to tell you much about the telephone receiver for it deserves a whole letter all to itself. You know that a magnet attracts iron. Suppose you wind a coil of insulated wire around a bar magnet or put the magnet inside such a coil as in Fig. 58. Send a stream of electrons through the turns of the coil–a steady stream such as comes from the battery shown in 133the figure. The strength of the magnet is altered. For one direction of the electron stream through the coil the magnet is stronger. For the opposite direction of current the magnet will be weaker.

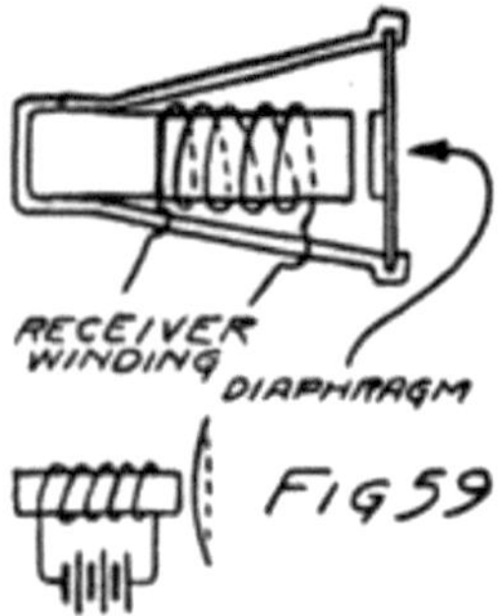

Fig. 59 shows a simple design of telephone receiver. It is formed by a bar magnet, a coil about it through which a current can flow, and a thin disc of iron. The iron disc, or diaphragm, is held at its edges so that it cannot move as a whole toward the magnet. The center can move, however, and so the diaphragm is bowed out in the form shown in the smaller sketch.

Now connect a battery to the receiver winding and allow a steady stream of electrons to flow. The magnet will be either strengthened or weakened. Suppose the stream of electrons is in the direction to make it stronger–I'll give you the rule later. Then the diaphragm is bowed out still more. If we open the battery circuit and so stop the stream of electrons the diaphragm will fly back to its original position, for it is elastic. The effect is very much that of pushing in the bottom of a tin pan and letting it fly back when you remove your hand.

Next reverse the battery. The magnet does not pull as hard as it would if there were no current. The diaphragm is therefore not bowed out so much.

Suppose that instead of reversing the current by reversing the battery we arrange to send an alternating 134current through the coil. That will have the same effect. For one direction of current flow, the diaphragm is attracted still more by the magnet but for the other direction it is not attracted as much. The result is that the center of the diaphragm moves back and forth during one complete cycle of the alternating current in the coil.

The diaphragm vibrates back and forth in tune with the alternating current in the receiver winding. As it moves away from the magnet it pushes ahead of it the neighboring molecules of air. These molecules then crowd and push the molecules of air which are just a little further away from the diaphragm. These in turn push against those beyond them and so a push or shove is sent out by the diaphragm from molecule to molecule until perhaps it reaches your ear. When the molecules of air next your ear receive the push they in turn push against your eardrum.

In the meantime what has happened? The current in the telephone receiver has reversed its direction. The diaphragm is now pulled toward the magnet and the adjacent molecules of air have even more room than they had before. So they stop crowding each other and follow the diaphragm in the other direction. The molecules of air just beyond these, on the way toward your ear, need crowd no longer and they also move back. Of course, they go even farther than their old positions for there is now more room on the other side. That same thing happens all along the line until the air molecules next your ear start back and give your eardrum a chance 135to expand outward. As they move away they make a little vacuum there and the eardrum puffs out.

That goes on over and over again just as often as the alternating current passes through one cycle of values. And you, unless you are thinking particularly of the scientific explanations, say that you "hear a musical note." As a matter of fact if we increase the frequency of the alternating current you will say that the "pitch" of the note has been increased or that you hear a note higher in the musical scale.

If we started with a very low-frequency alternating current, say one of fifteen or twenty cycles per second, you wouldn't say you heard a note at all. You would hear a sort of a rumble. If we should gradually increase the frequency of the alternating current you would find that about sixty or perhaps a hundred cycles a second would give you the impression of a musical note. As the frequency is made still larger you have merely the impression of a higher-pitched note until we get up into the thousands of cycles a second. Then, perhaps about twenty-thousand cycles a second, you find you hear only a little sound like wind or like steam escaping slowly from a jet or through a leak. A few thousand cycles more each second and you don't hear anything at all.

You know that for radio-transmitting stations we use audion oscillators which are producing alternating currents with frequencies of several hundred-thousand cycles per second. It certainly wouldn't do any good to connect a telephone receiver in the 136antenna circuit at the receiving station as in Fig. 60. We couldn't hear so high pitched a note.

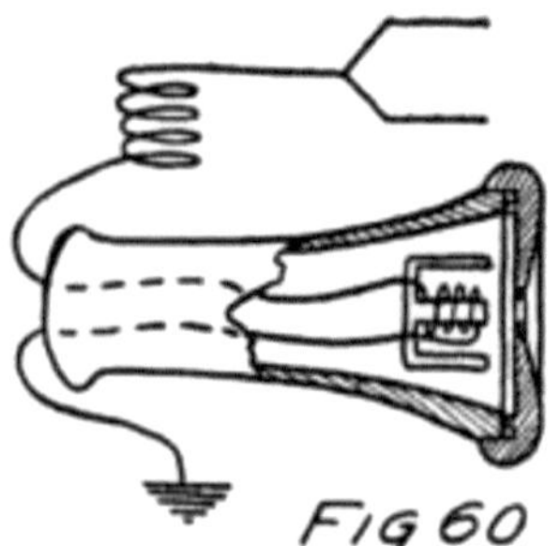

Even if we could, there are several reasons why the telephone receiver wouldn't work at such high frequencies. The first is that the diaphragm can't be moved so fast. It has some inertia, you know, that is, some unwillingness to get started. If you try to start it in one direction and, before you really get it going, change your mind and try to make it go in the other direction, it simply isn't going to go at all. So even if there is an alternating current in the coil around the magnet there will not be any corresponding vibration of the diaphragm if the frequency is very high, certainly not if it is above about 20,000 cycles a second.

The other reason is that there will only be a very feeble current in the coil anyway, no matter what you do, if the frequency is high. You remember that the electrons in a coil are sort of banded together and each has an effect on all the others which can move in parallel paths. The result is that they have a great unwillingness to get started and an equal unwillingness to stop. Their unwillingness is much more than if the wire was long and straight. It is also made very much greater by the presence of the iron core. An alternating e. m. f. of high frequency hardly gets the electrons started at all before it's 137time to get them going in the opposite direction. There is very little movement to the electrons and hence only a very small current in the coil if the frequency is high.

If you want a rule for it you can remember that the higher the frequency of an alternating e. m. f. the smaller the electron stream which it can set oscillating in a given coil. Of course, we might make the e. m. f. stronger, that is pull and shove the electrons harder, but unless the coil has a very small inductance or unless the frequency is very low we should have to use an e. m. f. of enormous strength to get any appreciable current.

Condensers are just the other way in their action. If there is a condenser in a circuit, where an alternating e. m. f. is active, there is lots of trouble if the frequency is low. If, however, the frequency is high the same-sized current can be maintained by a smaller e. m. f. than if the frequency is low. You see, when the frequency is high the electrons hardly get into the waiting-room of the condenser before it is time for them to turn around and go toward the other room. Unless there is a large current, there are not enough electrons crowded together in the waiting-room to push back very hard on the next one to be sent along by the e. m. f. Because the electrons do not push back very hard a small e. m. f. can drive them back and forth.

Ordinarily we say that a condenser impedes an alternating current less and less the higher is the frequency of the current. And as to inductances, we say that an inductance impedes an alternating current 138more and more the higher is the frequency.

Now we are ready to study the receiving circuit of Fig. 54. I showed you in Fig. 57 how the current through, the tube will vary as time goes on. It increases and decreases with the frequency of the current in the antenna of the distant transmitting station. We have a picture, or graph, as we say, of how this plate current varies. It will be necessary to study that carefully and to resolve it into its components, that is to separate it into parts, which, added together again will give the whole. To show you what I mean I am going to treat first a very simple case involving money.

Suppose a boy was started by his father with 50 cents of spending money. He spends that and runs 50 cents in debt. The next day his father gives him a dollar. Half of this he has to spend to pay up his yesterday's indebtedness. This he does at once and that leaves him 50 cents ahead. But again he buys something for a dollar and so runs 50 cents in debt. Day after day this cycle is repeated. We can show what happens by the curve of Fig. 61a.

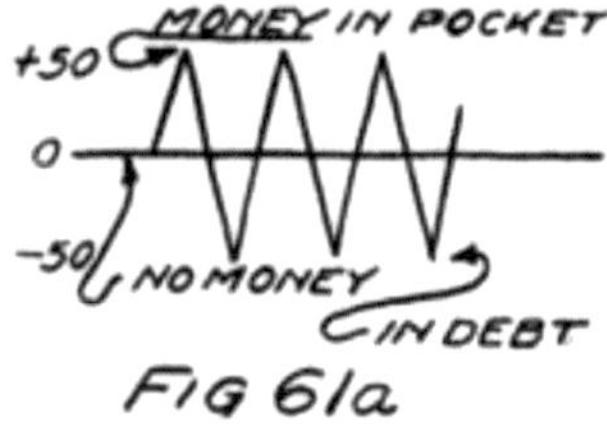

FIG 61a

On the other hand, suppose he already had 60 cents which, he was saving for some special purpose. This he doesn't touch, preferring to run into debt each day and to pay up the next, as shown in Fig. 61a. Then we would represent the story of this 60 cents by the graph of Fig. 61b.

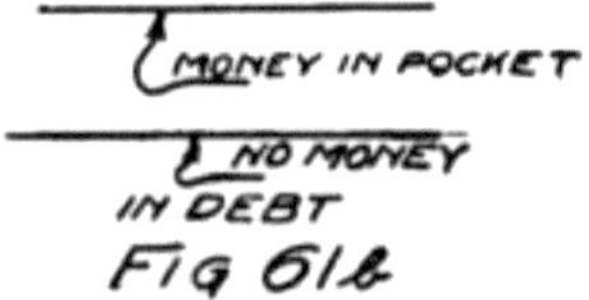

FIG 61b

139Now suppose that instead of going in debt each day he uses part of this 60 cents. Each day after the first his father gives him a dollar, just as before. He starts then with 60 cents as shown in Fig. 61c, increases in wealth to $1.10, then spends $1.00, bringing his

132

funds down to 10 cents. Then he receives $1.00 from his father and the process is repeated cyclically.

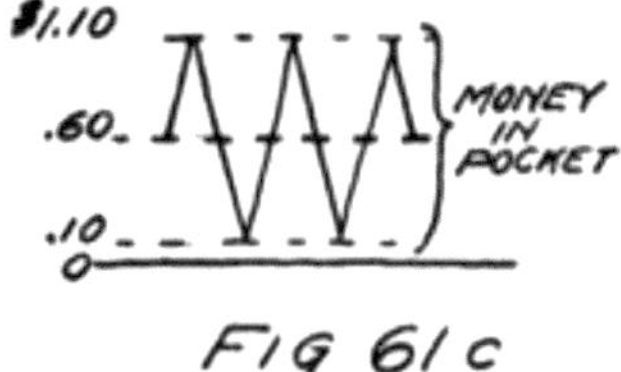

FIG 61c

If you saw the graph of Fig. 61c you would be able to say that, whatever he actually did, the effect was the same as if he had two pockets, in one of which he kept 60 cents all the time as shown in Fig. 61b. In his other pocket he either had money or he was in debt as shown in Fig. 61a. If you did that you would be resolving the money changes of Fig. 61c into the two components of Figs. 61a and b.

That is what I want you to do with the curve of Fig. 57 which I am reproducing here, redrawn as Fig. 62a. You see it is really the result of adding together the two curves of Figs. 62b and c, which are shown on the following page.

FIG 62a

We can think, therefore, of the current in the plate circuit as if it were two currents added together, that is, two electron streams passing through the same 140wire. One stream is steady and the other alternates.

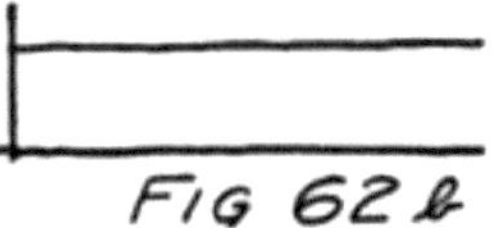

FIG 62b

Now look again at the diagram of our receiving set which I am reproducing as Fig. 63. When the signal is incoming there flow in the plate circuit two streams of electrons, one steady and of a value in mil-amperes corresponding to that of the graph in Fig. 62b, and the other alternating as shown in Fig. 62c.

The steady stream of electrons will have no more difficulty in getting through the coiled wire of the receiver than it would through the same amount of straight wire. On the other hand it cannot pass the gap of the condenser.

The alternating-current component can't get along in the coil because its frequency is so high that the coil impedes the motion of the electrons so much as practically to stop them. On the other hand these electrons can easily run into the waiting-room offered by the condenser and then run out again as soon as it is time.

FIG 62c

When the current in the plate circuit is large all the electrons which aren't needed for the steady stream through the telephone receiver run into one plate of the condenser. Of course, at that same instant an equal number leave the other plate and start off toward the B-battery and the filament. An instant later, when the current in the plate circuit is small, electrons start to come out of the 141plate and to join the stream through the receiver so that this stream is kept steady.

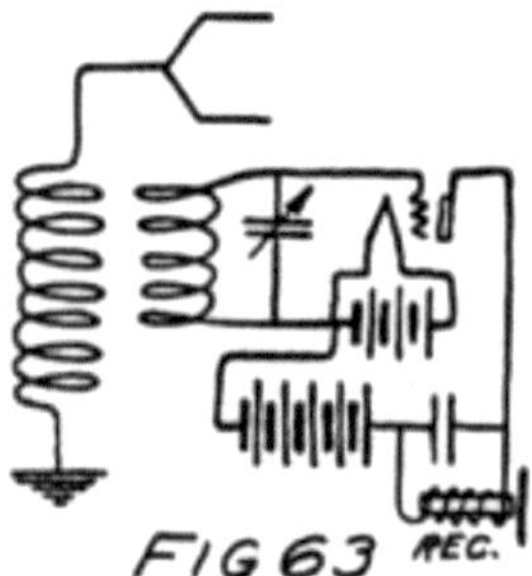

FIG 63

This steady stream of electrons, which is passing through the receiver winding, is larger than it would be if there was no incoming radio signal. The result is a stronger pull on the diaphragm of the receiver. The moment the signal starts this diaphragm is pulled over toward the magnet and it stays pulled over as long as the signal lasts. When the signal ceases it flies back. We would hear then a click when the signal started and another when it stopped.

If we wanted to distinguish dots from dashes this wouldn't be at all satisfactory. So in the next letter I'll show you what sort of changes we can make in the apparatus. To understand what effect these changes will have you need, however, to understand pretty well most of this letter.

142 LETTER 15
RADIO-TELEPHONY

DEAR LAD:

Before we start on the important subject matter of this letter let us make a short review of the preceding two letters.

An oscillating audion at the transmitting station produces an effect on the plate current of the detector audion at the receiving station. There is impressed upon the grid of the detector an alternating e. m. f. which has the same frequency as the alternating current which is being produced at the sending station. While this e. m. f. is active, and of course it is active only while the sending key is held down, there is more current through the winding of the telephone receiver and its diaphragm is consequently pulled closer to its magnet.

What will happen if the e. m. f. which is active on the grid of the detector is made stronger or weaker? The pull on the receiver diaphragm will be stronger or weaker and the diaphragm will have to move accordingly. If the pull is weaker the elasticity of the iron will move the diaphragm away from the magnet, but if the pull is stronger the diaphragm will be moved toward the magnet.

Every time the diaphragm moves it affects the air in the immediate neighborhood of itself and that air 143in turn affects the air farther away and so the ear of the listener. Therefore if there are changes in the intensity or strength of the incoming signal there are going to be corresponding motions of the receiver diaphragm. And something to listen, too, if these changes are frequent enough but not so frequent that the receiver diaphragm has difficulty in following them.

There are many ways of affecting the strength of the incoming signal. Suppose, for example, that we arrange to decrease the current in the antenna of the transmitting station. That will mean a weaker signal and a smaller increase in current through the winding of the telephone receiver at the other station. On the other hand if the signal strength is increased there is more current in this winding.

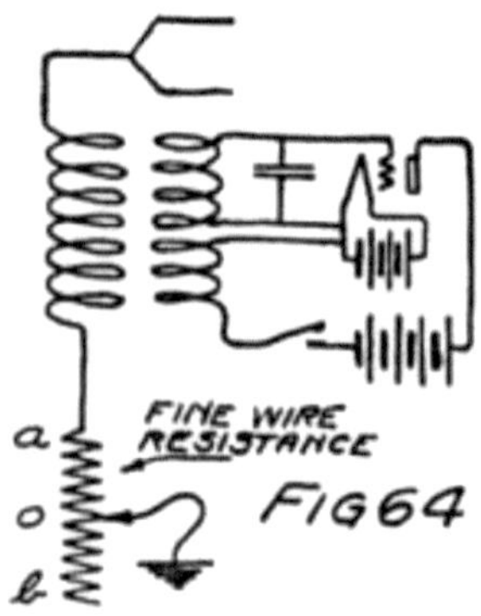

Suppose we connect a fine wire in the antenna circuit as in Fig. 64 and have a sliding contact as shown. Suppose that when we depress the switch in the oscillator circuit and so start the oscillations that the sliding contact is at *o* as shown. Corresponding to that strength of signal there is a certain value of current through the receiver winding at the other station. Now let us move the slider, first to *a* and then back to *b* and so on, back and forth. You see what will happen. We alternately make the current in the antenna 144 larger and smaller than it originally was. When the slider is at *b* there is more of the fine wire in series with the antenna, hence more re- sistance to the oscillations of the electrons, and hence a smaller os- cillating stream of electrons. That means a weaker outgoing signal. When the slider is at *a* there is less resistance in the antenna circuit and a larger alternating current.

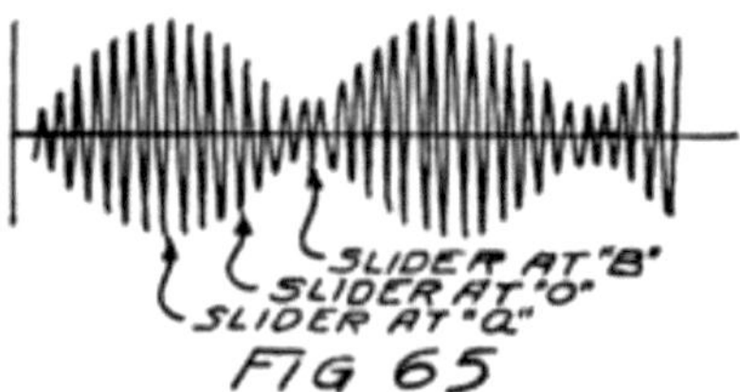

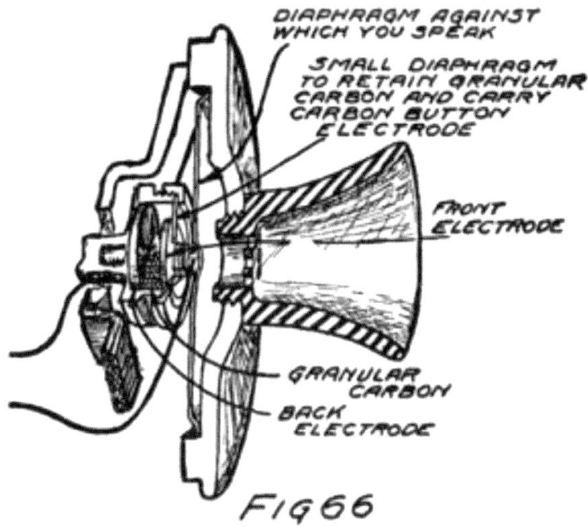

A picture of what happens would be like that of Fig. 65. The signal varies in intensity, therefore, becoming larger and smaller alternately. That means the voltage impressed on the grid of the detector is alternately larger and smaller. And hence the stream of electrons through the winding of the telephone receiver is alternately larger and smaller. 145And that means that the diaphragm moves back and forth in just the time it takes to move the slider back and forth.

Instead of the slider we might use a little cup almost full of grains of carbon. The carbon grains lie between two flat discs of carbon. One of these discs is held fixed. The other is connected to the center of a thin diaphragm of steel and moves back and forth as this diaphragm is moved. The whole thing makes a telephone transmitter such as you have often talked to.

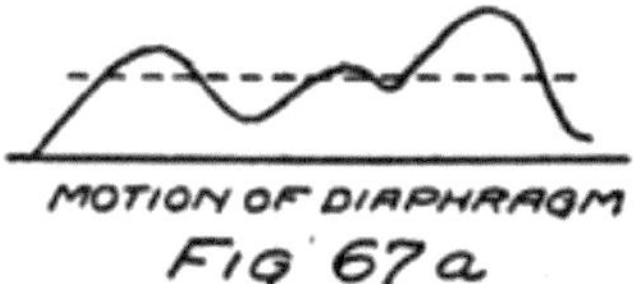

Wires connect to the carbon discs as shown in Fig. 66. A stream of electrons can flow through the wires and from grain to grain through the "carbon button," as we call it. The electrons have less difficulty if the grains are compressed, that is the button then offers less resistance to the flow of current. If the diaphragm moves back, allowing the grains to have more room, the electron stream is smaller and we say the button is offering more resistance to the current.

You can see what happens. Suppose some one talks into the transmitter and makes its diaphragm go back and forth as shown in Fig. 67a. Then the current in the antenna varies, being greater or less, depending upon whether the button offers less or 146more resistance. The corresponding variations in the antenna current are shown in Fig. 67b.

In the antenna at the receiving station there are corresponding variations in the strength of the signal and hence corresponding variations in the strength of the current through the telephone receiver. I shall show graphically what happens in Fig. 68. You see that the telephone receiver diaphragm does just the same motions as does the transmitter diaphragm. That means that the molecules of air near the receiver diaphragm are going through just the same kind of motions as are those near the transmitter diaphragm. When these air molecules affect your eardrum you hear just what you would have heard if you had been right there beside the transmitter.

That's one way of making a radio-telephone. It is not a very efficient method but it has been used in the past. Before we look at any of the more recent methods we can draw some general ideas from this method and learn some words that are used almost always in speaking of radio-telephones.

In any system of radio-telephony you will always find that there is produced at the transmitting station a high-frequency alternating current and that this current flows in a tuned circuit one part of which is the condenser formed by the antenna and the ground (or something which acts like a ground). This high-frequency current, or radio-current, as we usually say, is varied in its strength. It is varied in conformity with the human voice. If the human 147voice speaking into the transmitter is low pitched there are slow variations in the intensity of the radio current. If the voice is high pitched there are more rapid variations in the strength of the radio-frequency current. That is why we say the radio-current is "modulated" by the human voice.

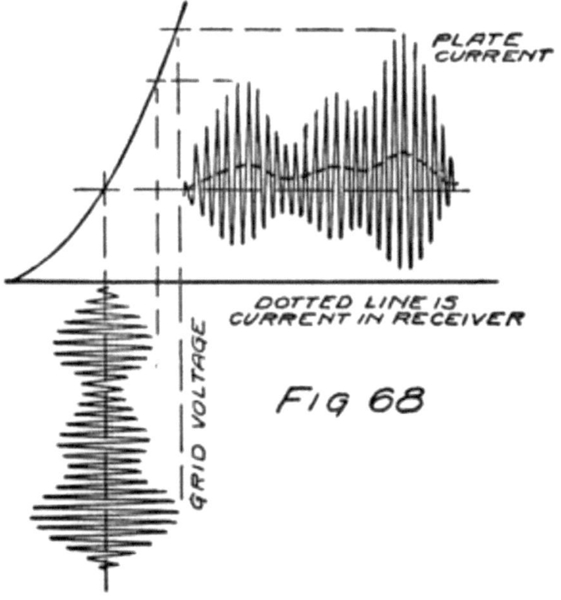

The signal which radiates out from the transmitting antenna carries all the little variations in pitch and loudness of the human voice. When this signal reaches the distant antenna it establishes in that antenna circuit a current of high frequency which has just the same variations as did the current in the antenna at the sending station. The human voice isn't there. It is not transmitted. It did its work at the sending station by modulating the radio-signal, 148"modulating the carrier current," as we sometimes say. But there is speech significance hidden in the variations in strength of the received signal.

If a telephone-receiver diaphragm can be made to vibrate in accordance with the variations in signal intensity then the air adjacent to that diaphragm will be set into vibration and these vibrations will be just like those which the human voice set up in the air molecules near the mouth of the speaker. All the different systems of receiving radio-telephone signals are merely different methods of getting a current which will affect the telephone receiver in conformity with the variations in signal strength. Getting such a current is called "detecting." There are many different kinds of detectors but the vacuum tube is much to be preferred.

The cheapest detector, but not the most sensitive, is the crystal. If you understand how the audion works as a detector you will have no difficulty in understanding the crystal detector.

The crystal detector consists of some mineral crystal and a fine-wire point, usually platinum. Crystals are peculiar things. Like everything else they are made of molecules and these molecules of atoms. The atoms are made of electrons grouped around nuclei which, in turn, are formed by close groupings of protons and electrons. The great difference between crystals and substances which are not crystalline, that is, substances which don't have a special natural shape, is this: In crystals the molecules and atoms are all arranged in some orderly manner. 149In other substances, substances without special form, amorphous substances, as we call them, the molecules are just grouped together in a haphazard way.

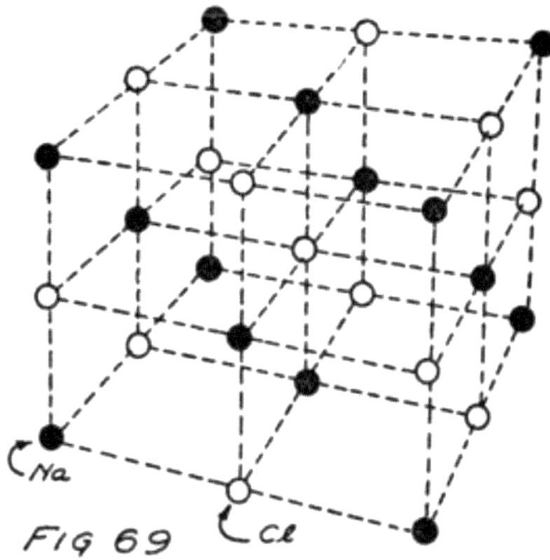

FIG 69

For some crystals we know very closely indeed how their molecules or rather their individual atoms are arranged. Sometime you may wish to read how this was found out by the use of X-rays. [6] Take the crystal of common salt for example. That is well known. Each molecule of salt is formed by an atom of sodium and one of chlorine. In a crystal of salt the molecules are grouped together so that a sodium atom always has chlorine atoms on every side of it, and the other way around, of course.

Suppose you took a lot of wood dumb-bells and painted one of the balls of each dumb-bell black to stand for a sodium atom, leaving the other unpainted 150to stand for a chlorine atom. Now try to pile them up so that above and below each black ball, to the right and left of it, and also in front and behind it, there shall be a white ball. The pile which you would probably get would look like that of Fig. 69. I have omitted the gripping part of each dumbell because I don't believe it is there. In my picture each circle represents the nucleus of an atom. I haven't attempted to show the planetary electrons. Other crystals have more complex arrangements for piling up their molecules.

Now suppose we put two different kinds of substances close together, that is, make contact between them. How their electrons will behave will depend entirely upon what the atoms are and how they are piled up. Some very curious effects can be obtained.

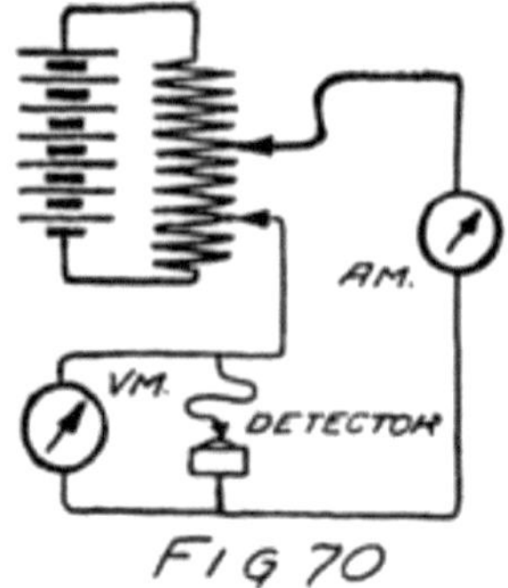

FIG 70

The one which interests us at present is that across the contact points of some combinations of substances it is easier to get a stream of electrons to flow one way than the other. The contact doesn't have the same resistance in the two directions. Usually also the resistance depends upon what voltage we are applying to force the electron stream across the point of contact.

The one way to find out is to take the voltage-current characteristic of the combination. To do so we use the same general method as we did for the audion. 151And when we get through we plot another curve and call it, for example, a "platinum-galena characteristic." Fig. 70 shows the set-up for making the measurements. There is a group of batteries arranged so that we can vary the e. m. f. applied across the contact point of the crystal and platinum. A voltmeter shows the value of this e. m. f. and an ammeter tells the strength of the electron stream. Each time we move the slider we get a new pair of values for volts and amperes. As a matter of fact we don't get amperes or even mil-amperes; we get millionths of an ampere or "microamperes," as we say. We can plot the pairs of values which we measure and make a curve like that of Fig. 71.

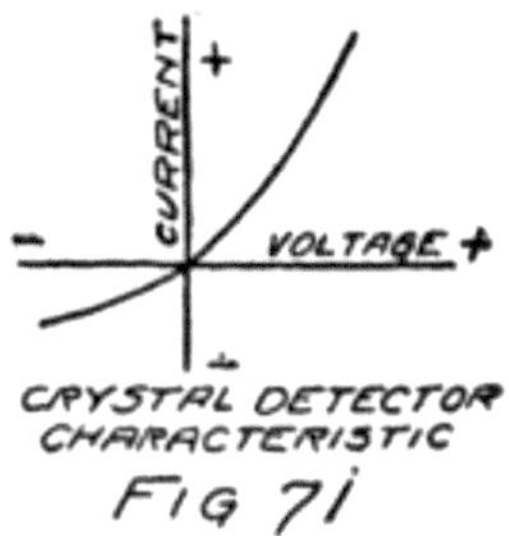

When the voltage across the contact is reversed, of course, the current reverses. Part of the curve looks something like the lower part of an audion characteristic.

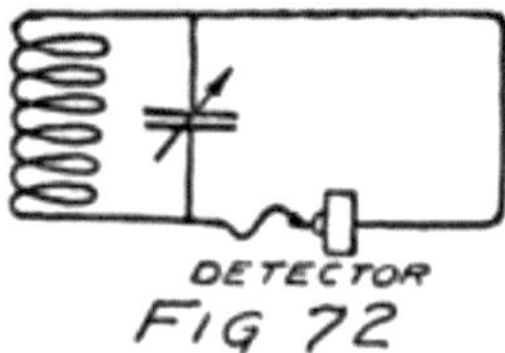

Now connect this crystal in a receiving circuit as in Fig. 72. We use an antenna just as we did for the audion and we tune the antenna circuit to the frequency of the incoming signal. The receiving circuit is coupled to the antenna circuit and is tuned to the same frequency. Whatever voltage there may be across the condenser of this circuit is applied to the crystal detector. We haven't put the telephone 152receiver in the circuit yet. I want to wait until you have seen what the crystal does when an alternating voltage is applied to it.

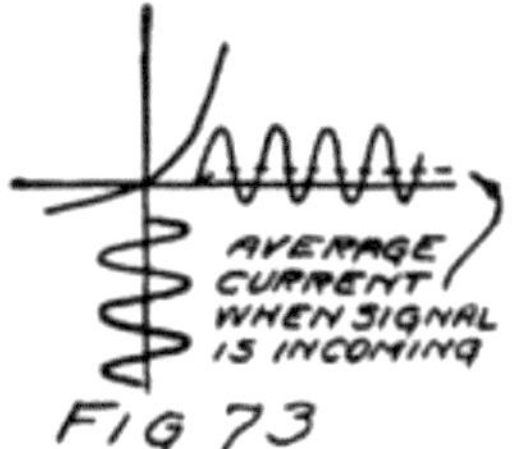

We can draw a familiar form of sketch as in Fig. 73 to show how the current in the crystal varies. You see that there flows through the crystal a current very much like that of Fig. 62a. And you know that such a current is really equivalent to two electron streams, one steady and the other alternating. The crystal detector gives us much the same sort of a current as does the vacuum tube detector of Fig. 54. The current isn't anywhere near as large, however, for it is microamperes instead of mil-amperes.

Our crystal detector produces the same results so far as giving us a steady component of current to send through a telephone receiver. So we can connect a receiver in series with the crystal as shown in Fig. 74. Because the receiver would offer a large impedance to the high-frequency current, that is, seriously impede and so reduce the high-frequency current, we connect a condenser around the receiver.

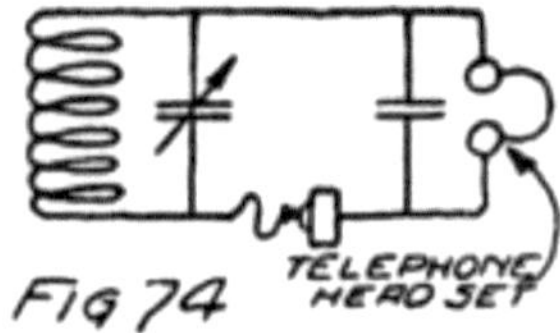

There is a simple crystal detector circuit. If the signal intensity varies then the current which passes through the receiver will vary. If these variations are caused by a human voice at the sending station the crystal will permit one to hear 153from the telephone receiver what the speaker is saying. That is just what the audion detector does very many times better.

In the letter on how to experiment you'll find details as to the construction of a crystal-detector set. Excellent instructions for an inexpensive set are contained in Bull. No. 120 of the Bureau of Standards. A copy can be obtained by sending ten cents to the Commissioner of Public Documents, Washington, D. C.

[6]

Cf. "Within the Atom," Chapter X.

154 LETTER 16
THE HUMAN VOICE

The radio-telephone does not transmit the human voice. It reproduces near the ears of the listener similar motions of the air molecules and hence causes in the ears of the listener the same sensations of sound as if he were listening directly to the speaker. This reproduction takes place almost instantaneously so great is the speed with which the electrical effects travel outward from the sending antenna. If you wish to understand radio-telephony you must know something of the mechanism by which the voice is produced and something of the peculiar or characteristic properties of voice sounds.

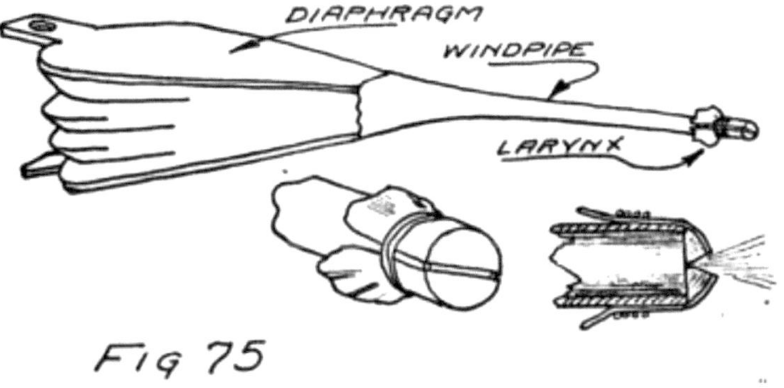

FIG 75

The human voice is produced by a sort of organ pipe. Imagine a long pipe connected at one end to a pair of fire-bellows, and closed at the other end by two stretched sheets of rubber. Fig. 75 is a sketch of 155what I mean. Corresponding to the bellows there is the human diaphragm, the muscular membrane separating the thorax and abdomen, which expands or contracts as one breathes. Corresponding to the pipe is the windpipe. Corresponding to the two stretched pieces of rubber are the vocal cords, L and R, shown in cross section in Fig. 77. They are part of the larynx and do not show in Fig. 76 (Pl. viii) which shows the wind pipe and an outside view of the larynx.

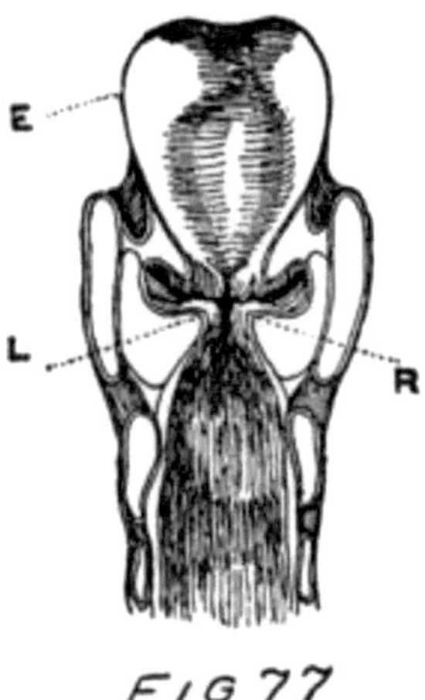

FIG 77

When the sides of the bellows are squeezed together the air molecules within are crowded closer together and the air is compressed. The greater the compression the greater, of course, is the pressure with which the enclosed air seeks to escape. That it can do only by lifting up, that is by blowing out, the two elastic strips which close the end of the pipe.

The air pressure, therefore, rises until it is sufficient to push aside the elastic membranes or vocal cords and thus to permit some of the air to escape. It doesn't force the membranes far apart, just enough to let some air out. But the moment some air has escaped there isn't so much inside and the pressure is reduced just as in the case of an automobile tire from which you let the air escape. What is the result? The membranes fly back again and 156close the opening of the pipe. What got out, then, was just a little puff of air.

The bellows are working all the while, however, and so the space available for the remaining air soon again becomes so crowded with air molecules that the pressure is again sufficient to open the membranes. Another puff of air escapes.

This happens over and over again while one is speaking or singing. Hundreds of times a second the vocal cords vibrate back and forth. The frequency with which they do so determines the note or pitch of the speaker's voice.

148

What determines the significance of the sounds which he utters? This is a most interesting question and one deserving of much more time than I propose to devote to it. To give you enough of an answer for your study of radio-telephony I am going to tell you first about vibrating strings for they are easier to picture than membranes like the vocal cords.

Suppose you have a stretched string, a piece of rubber band or a wire will do. You pluck it, that is pull it to one side. When you let go it flies back. Because it has inertia [7] it doesn't stop when it gets to its old position but goes on through until it bows out almost as far on the other side.

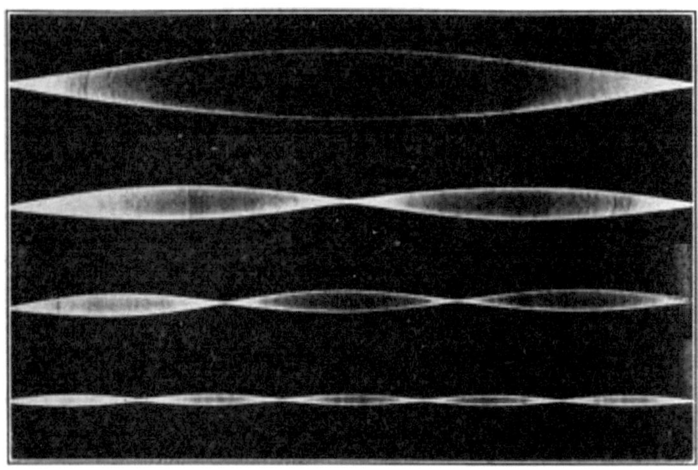

FIG 79

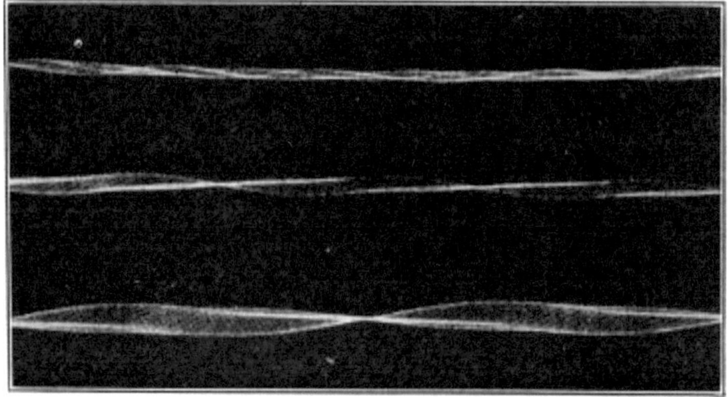

FIG 80

Pl. VII.–Photographs of Vibrating Strings.

157It took some work to pluck this string, not much perhaps; but all the work which you did in deforming it, goes to the string and becomes its energy, its ability to do work. This work it does in pushing the air molecules ahead of it as it vibrates. In this way it uses up its energy and so finally comes again to rest. Its vibrations "damp out," as we say, that is die down. Each swing carries it a smaller distance away from its original position. We say that the "amplitude," meaning the size, of its vibration decreases. The frequency does not. It takes just as long for a small-sized vibration as for the larger. Of course, for the vibration of large amplitude the string must move faster but it has to move farther so that the time required for a vibration is not changed.

First the string crowds against each other the air molecules which are in its way and so leads to crowding further away, just as fast as these molecules can pass along the shove they are receiving. That takes place at the rate of about 1100 feet a second. When the string swings back it pushes away the molecules which are behind it and so lets those that were being crowded follow it. You know that they will. Air molecules will always go where there is the least crowding. Following the shove, therefore, there is a chance for the molecules to move back and even to occupy more room than they had originally.

The news of this travels out from the string just as fast as did the news of the crowding. As fast as molecules are able they move back and so make more room for their neighbors who are farther away; and these in turn move back.

Do you want a picture of it? Imagine a great crowd of people and at the center some one with authority. The crowd is the molecules of air and the 158one with authority is one of the molecules of the string which has energy. Whatever this molecule of the string says is repeated by each member of the crowd to his neighbor next farther away. First the string says: "Go back" and each molecule acts as soon as he gets the word. And then the string says: "Come on" and each molecule of air obeys as soon as the command reaches him. Over and over this happens, as many times a second as the string makes complete vibrations.

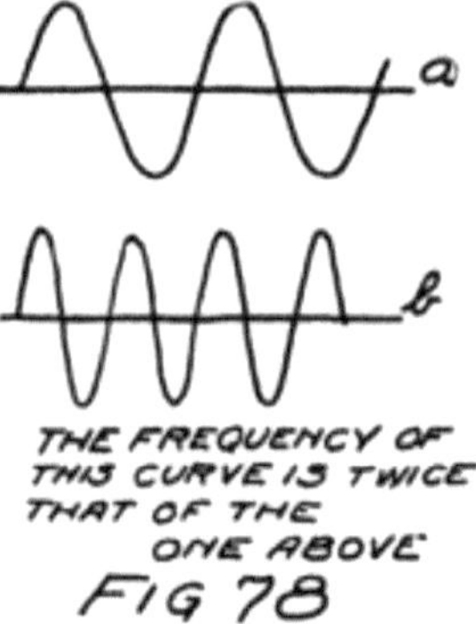

FIG 78

If we should make a picture of the various positions of one of these air molecules much as we pictured "Brownie" in Letter 9 it would appear as in Fig. 78a where the central line represents the ordinary position of the molecule.

That's exactly the picture also of the successive positions of an electron in a circuit which is "carrying an alternating current." First it moves in one direction along the wire and then back in the opposite direction. The electron next to it does the same thing almost immediately for it does not take anywhere near as long for such an effect to pass through a crowd of electrons. If we make the string vibrate twice as fast, that is, have twice the frequency, the story of an adjacent particle of air will be as in Fig. 78b. Unless we tighten the string, however, we can't make it vibrate 159as a whole and do it twice as fast. We can make it vibrate in two parts or even in more parts, as shown in Fig. 79 of Pl. VII. When it vibrates as a whole, its frequency is the lowest possible, the fundamental frequency as we say. When it vibrates in two parts each part of the string makes twice as many vibrations each second. So do the adjacent molecules of air and so does the eardrum of a listener.

The result is that the listener hears a note of twice the frequency that he did when the string was vibrating as a whole. He says he hears the "octave" of the note he heard first. If the string vibrates in three parts and gives a note of three times the frequency the listener hears a note two octaves above the "fundamental note" of which the string is capable.

152

It is entirely possible, however, for a string to vibrate simultaneously in a number of ways and so to give not only its fundamental note but several others at the same time. The photographs [8] of Fig. 80 of Pl. VII illustrate this possibility.

What happens then to the molecules of air which are adjacent to the vibrating string? They must perform quite complex vibrations for they are called upon to move back and forth just as if there were several strings all trying to push them with different frequencies of vibration. Look again at the pictures, of Fig. 80 and see that each might just as well be 160the picture of several strings placed close together, each vibrating in a different way. Each of the strings has a different frequency of vibration and a different maximum amplitude, that is, greatest size of swing away from its straight position.

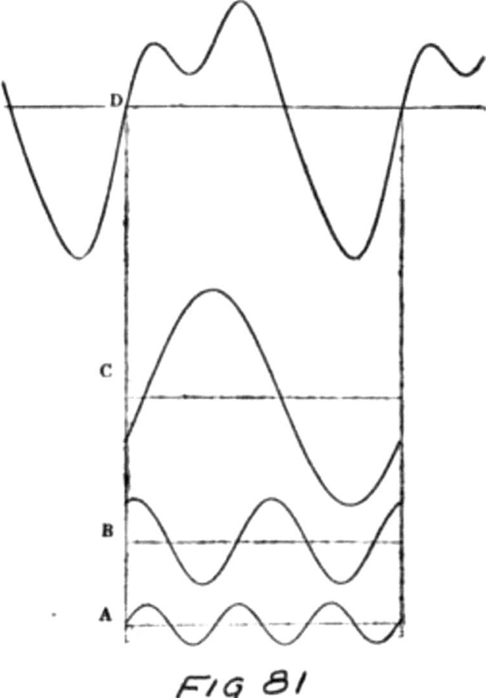

FIG 81

Suppose instead of a single string acting upon the adjacent molecules we had three strings. Suppose the first would make a nearby

153

molecule move as in Fig. 81A, the second as in Fig. 81B, and the third as in Fig. 81C. It is quite evident that the molecule can satisfy all three if it will vibrate as in Fig. 81D.

161Now take it the other way around. Suppose we had a picture of the motion of a molecule and that it was not simple like those shown in Fig. 78 but was complex like that of Fig. 81D. We could say that this complex motion was made up of three parts, that is, had three component simple motions, each represented by one of the three other graphs of Fig. 81. That means we can resolve any complex vibratory motion into component motions which are simple.

It means more than that. It means that the vibrating string which makes the neighboring molecules of air behave as shown in Fig. 81D is really acting like three strings and is producing simultaneously three pure musical notes.

Now suppose we had two different strings, say a piano string in the piano and a violin string on its proper mounting. Suppose we played both instruments and some musician told us they were in tune. What would he mean? He would mean that both strings vibrated with the same fundamental frequency.

They differ, however, in the other notes which they produce at the same time that they produce their fundamental notes. That is, they differ in the frequencies and amplitudes of these other component vibrations or "overtones" which are going on at the same time as their fundamental vibrations. It is this difference which lets us tell at once which instrument is being played.

That brings us to the main idea about musical sounds and about human speech. The pitch of any 162complex sound is the pitch of its fundamental or lowest sound; but the character of the complex sound depends upon all the overtones or "harmonics" which are being produced and upon their relative frequencies and amplitudes.

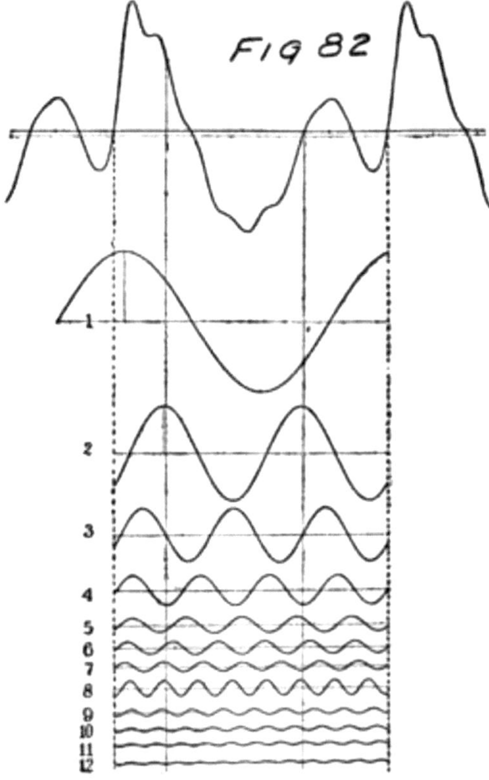

The organ pipe which ends in the larynx produces a very complex sound. I can't show you how complex but I'll show you in Fig. 82 the complicated motion of an air molecule which is vibrating as the result of being near an organ pipe. (Organ pipes differ–this is only one case.) You can see that 163there are a large number of pure notes of various intensities, that is, strengths, which go to make up the sound which a listener to this organ pipe would hear. The note from the human pipe is much more complex.

When one speaks there are little puffs of air escaping from his larynx. The vocal cords vibrate as I explained. And the molecules of air near the larynx are set into very complex vibrations. These transmit their vibrations to other molecules until those in the mouth are reached. In the mouth, however, something very important happens.

Did you ever sing or howl down a rain barrel or into a long pipe or hallway and hear the sound? It sounds just about the same no matter who does it. The reason is that the long column of air in the pipe or barrel is set into vibration and vibrates according to its own ideas of how fast to do it. It has a "natural frequency" of its own. If in your voice there is a note of just that frequency it will respond beautifully. In fact it "resonates," or sings back, when it hears this note.

The net result is that it emphasizes this note so much that you don't hear any of the other component notes of your voice–all you hear is the rain barrel. We say it reinforces one of the component notes of your voice and makes it louder.

That same thing happens in the mouth cavity of a speaker. The size and shape of the column of air in the mouth can be varied by the tongue and lip positions and so there are many different possibilities 164of resonance. Depending on lip and tongue, different frequencies of the complex sound which comes from the larynx are reinforced. You can see that for yourself from Fig. 83 which shows the tongue positions for three different vowel sounds. You can see also from Fig. 84, which shows the mouth positions for the different vowels, how the size and shape of the mouth cavity is changed to give different sounds. These figures are in Pl. VIII.

The pitch of the note need not change as every singer knows. You can try that also for yourself by singing the vowel sound of "ahh" and then changing the shape of your mouth so as to give the sound "ah–aw–ow–ou." The pitch of the note will not change because the fundamental stays the same. The speech significance of the sound, however, changes completely because the mouth cavity resonates to different ones of the higher notes which come from the larynx along with the fundamental note.

Now you can see what is necessary for telephonic transmission. Each and every component note which enters into human speech must be transmitted and accurately reproduced by the receiver. More than that, all the proportions must be kept just the same as in the original spoken sound. We usually say that there must be reproduced in the air at the receiver exactly the same "wave form" as is present in the air at the transmitter. If that isn't done the speech won't be natural and one cannot recognize voices although he may understand pretty well. If 165there is too much "distortion" of the wave form, that is if the relative intensities of the component notes of the voice are too much altered, then there may even be a loss of intelligibility so that the listener cannot understand what is being said.

What particular notes are in the human voice depends partly on the person who is speaking. You know that the fundamental of a bass voice is lower than that of a soprano. Besides the fundamental, however, there are a lot of higher notes always present. This is particularly true when the spoken sound is a consonant, like "s" or "f" or "v." The particular notes, which are present and are important, depend upon what sound one is saying.

Usually, however, we find that if we can transmit and reproduce exactly all the notes which lie between a frequency of about 200 cycles a second and one of about 2000 cycles a second the reproduced speech will be quite natural and very intelligible. For singing and for transmitting instrumental music it is necessary to transmit and reproduce still higher notes.

What you will have to look out for, therefore, in a receiving set is that it does not cut out some of the high notes which are necessary to give the sound its naturalness. You will also have to make sure that your apparatus does not distort, that is, does not receive and reproduce some notes or "voice frequencies" more efficiently than it docs some others which are equally necessary. For that reason when you buy a transformer or a telephone receiver it is 166well to ask for a characteristic curve of the apparatus which will show how the action varies as the frequency of the current is varied. The action or response should, of course, be practically the same at all the frequencies within the necessary part of the voice range.

[7]

Cf. Chap. VI of "The Realities of Modern Science."

[8]

My thanks are due to Professor D. C. Miller and to the Macmillan Company for permission to reproduce Figs. 79 to 83 inclusive from that interesting book, "The Science of Musical Sounds."

167 LETTER 17
GRID BATTERIES AND GRID CONDENSERS FOR DETECTORS

DEAR SON:

You remember the audion characteristics which I used in Figs. 55, 56 and 57 of Letter 14 to show you how an incoming signal will affect the current in the plate circuit. Look again at these figures and you will see that these characteristics all had the same general shape but that they differed in their positions with reference to the "main streets" of "zero volts" on the grid and "zero mil-amperes" in the plate circuit. Changing the voltage of the B-battery in the plate circuit changed the position of the characteristic. We might say that changing the B-battery shifted the curve with reference to the axis of zero volts on the grid.

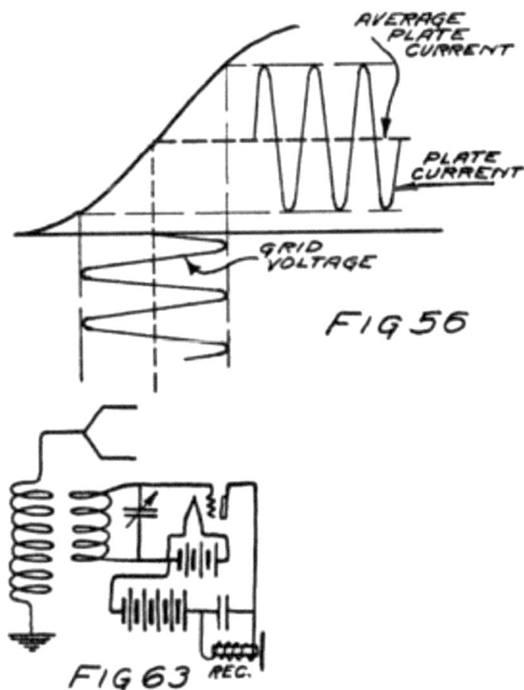

168In the case of the three characteristics which we are discussing the shift was made by changing the B-battery. Increasing B-voltage shifts characteristic to the left. It is possible, however, to produce such a shift by using a C-battery, that is, a battery in the grid circuit, which makes the grid permanently negative (or positive, depending upon how it is connected). This battery either helps or hinders the plate battery, and because of the strategic position of the grid right near the filament one volt applied to the grid produces as large an effect as would several volts in the plate battery. Usually, therefore, we arrange to shift the characteristic by using a C-battery.

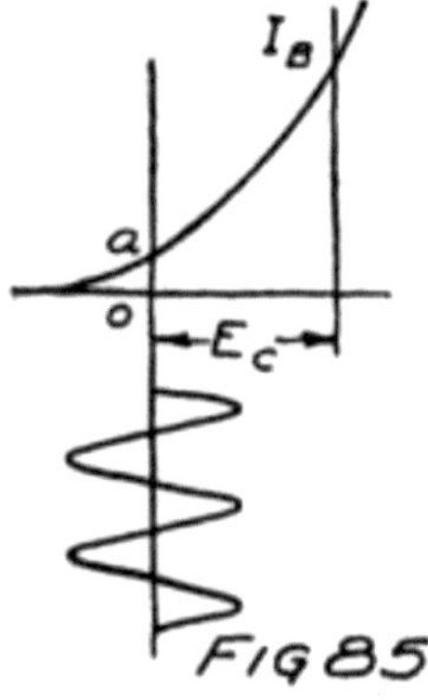

Suppose for example that we had an audion in the receiving circuit of Fig. 63 and that its characteristic under these conditions is given by Fig. 56. I've redrawn the figures to save your turning back. The audion will not act as a detector because an incoming signal will not change the average value of the current in the plate circuit. If, however, we connect a C-battery so as to make the grid negative, we can shift this characteristic so that the incoming signal will be detected. We have only to make the grid 169sufficiently negative to reduce the plate current to the value shown by the line *oa* in Fig. 85. Then the signal will be detected because, while it makes the plate current alternately larger and smaller than this value *oa*, it will result, on the average, in a higher value of the plate current.

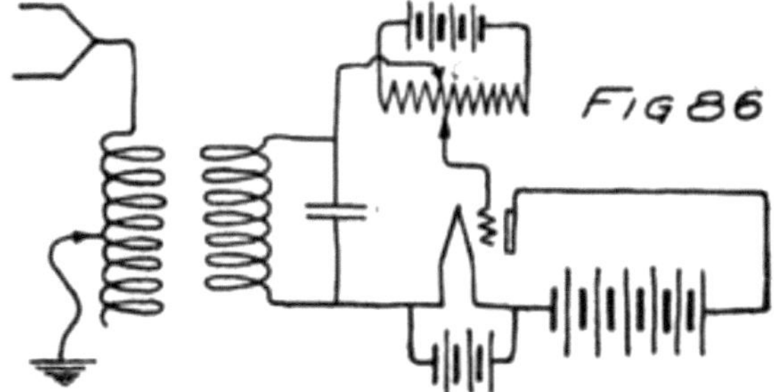

You see that what we have done is to arrange the point on the audion characteristic about which the tube is to work by properly choosing the value of the grid voltage E_C.

There is an important method of using an audion for a detector where we arrange to have the grid voltage change steadily, getting more and more negative all the time the signal is coming in. Before I tell how it is done I want to show you what will happen.

Suppose we start with an audion detector, for which the characteristic is that of Fig. 56, but arranged as in Fig. 86 to give the grid any potential which we wish. The batteries and slide wire resistance which are connected in the grid circuit are already familiar to you.

When the slider is set as shown in Fig. 86 the grid 170is at zero potential and we are at the point 1 of the characteristic shown in Fig. 87. Now imagine an incoming signal, as shown in that same figure, but suppose that as soon as the signal has stopped making the grid positive we shift the slider a little so that the C-battery makes the grid slightly negative. We have shifted the point on the characteristic about which the tube is being worked by the incoming signal from point 1 to point 2.

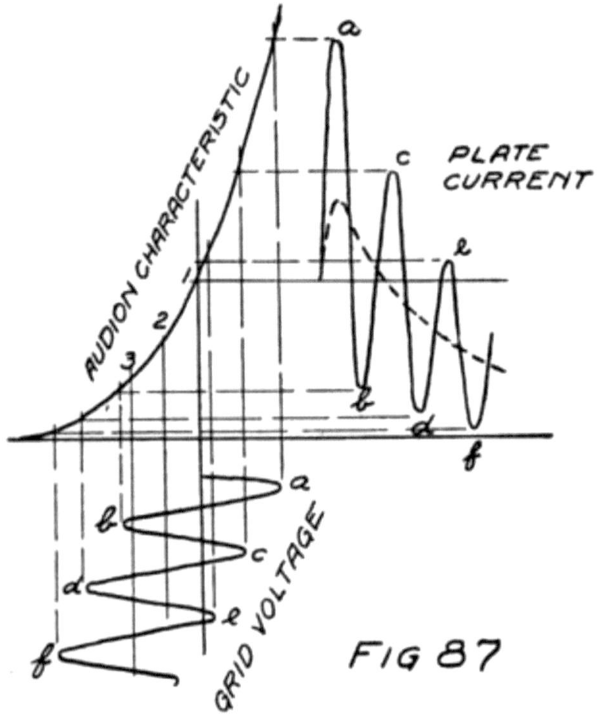

Every time the incoming signal makes one complete cycle of changes we shift the slider a little further and make the grid permanently more negative. You can see what happens. As the grid becomes more negative the current in the plate circuit decreases on the average. Finally, of course, the grid will become so negative that the current in the plate circuit will be reduced to zero. Under these conditions an incoming signal finally makes a large change in the plate current and hence in the current through the telephone.

The method of shifting a slider along, every time the incoming signal makes a complete cycle, is impossible to accomplish by hand if the frequency of the signal is high. It can be done automatically, however, no matter how high the frequency if we use a condenser in the grid circuit as shown in Fig. 88.

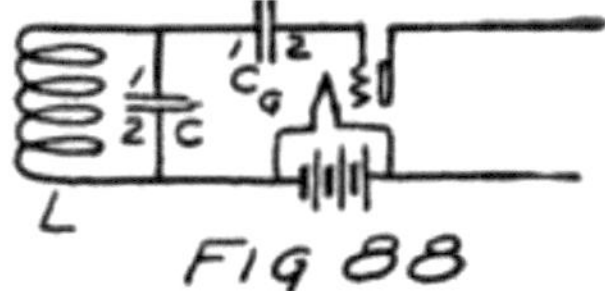

When the incoming signal starts a stream of electrons through the coil L of Fig. 88 and draws them away from plate 1 of the condenser C it is also drawing electrons away from the 1 plate of the condenser C_G which is in series with the grid. As electrons leave plate 1 of this condenser others rush away from the grid and enter plate 2. This means that the grid doesn't have its ordinary number of electrons and so is positive.

If the grid is positive it will be pleased to get electrons; and it can do so at once, for there are lots of electrons streaming past it on their way to the plate. While the grid is positive, therefore, there is a stream of electrons to it from the filament. Fig. 89 shows this current.

All this takes place during the first half-cycle of the incoming signal. During the next half-cycle electrons are sent into plate 1 of the condenser C and also into plate 1 of the grid condenser C_G. As electrons are forced into plate 1 of the grid condenser those in plate 2 of that condenser have to leave and go back to the grid where they came from. That is all right, but while they were away the grid got some electrons from the filament to take their places. The result is that the grid has now too many electrons, that is, it is negatively charged.

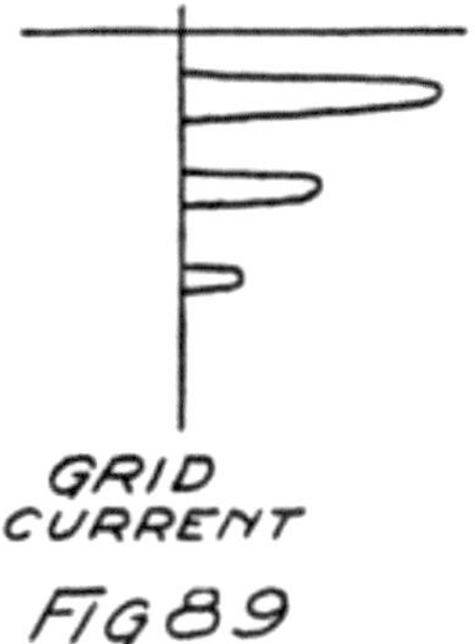

An instant later the signal e. m. f. reverses and calls electrons away from plate 1 of the grid condenser. Again electrons from the grid rush into plate 2 and again the grid is left without its proper number and so is positive. Again it receives electrons from the filament. The result is still more electrons in the part of the grid circuit which is formed by the grid, the plate 2 of the grid condenser and the connecting wire. These electrons can't get across the gap of the condenser C_G and they can't go back to the filament any other way. So there they are, trapped. Finally there are so many of these trapped electrons that the grid is so negative all the time as almost entirely to oppose the efforts of the plate to draw electrons away from the filament.

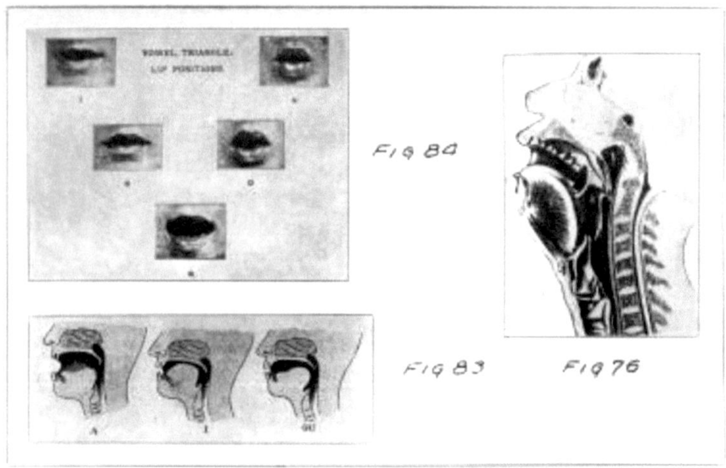

Pl. VIII.–To Illustrate the Mechanism for the Production of the Human Voice.

171 Then the plate current is reduced practically to zero.

That's the way to arrange an audion so that the incoming signal makes the largest possible change in plate current. We can tell if there is an incoming signal because it will "block" the tube, as we say. The plate-circuit current will be changed from its ordinary value to almost zero in the short time it takes for a few cycles of the incoming signal.

We can detect one signal that way, but only one because the first signal makes the grid permanently negative and blocks the tube so that there isn't any current in the plate circuit and can't be any. If we want to put the tube in condition to receive another signal we must allow these electrons, which originally came from the filament, to get out of their trapped position and go back to the filament.

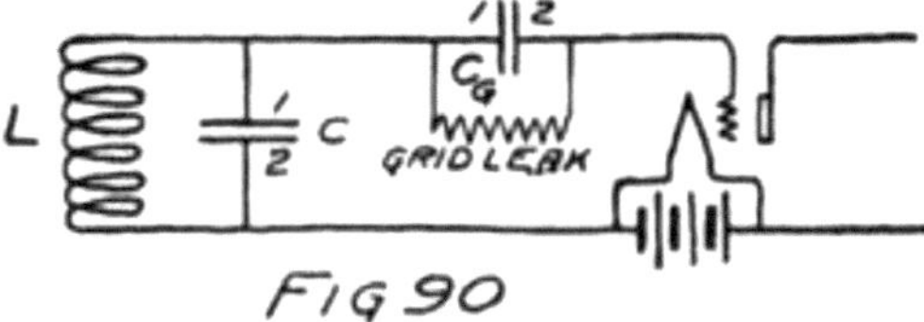

To do so we connect a very fine wire between plates 1 and 2 of the grid condenser. We call that wire a "grid-condenser leak" because it lets the electrons slip around past the gap. By using a very high resistance, we can make it so hard for the electrons to get around the gap that not many will do so while the signal is coming in. In that case we can leave the leak permanently across the condenser as shown in Fig. 90. Of course, the leak must offer so easy a path for the electrons that all the trapped 172electrons can get home between one incoming signal and the next.

One way of making a high resistance like this is to draw a heavy pencil line on a piece of paper, or better a line with India ink, that is ink made of fine ground particles of carbon. The leak should have a very high resistance, usually one or two million ohms if the condenser is about 0.002 microfarad. If it has a million ohms we say it has a "megohm" of resistance.

This method of detecting with a leaky grid-condenser and an audion is very efficient so far as telling the listener whether or not a signal is coming into his set. It is widely used in receiving radio-telephone signals although it is best adapted to receiving the telegraph signals from a spark set.

I don't propose to stop to tell you how a spark-set transmitter works. It is sufficient to say that when the key is depressed the set sends out radio signals at the rate usually of 1000 signals a second. Every time a signal reaches the receiving station the current in the telephone receiver is sudden reduced; and in the time between signals the leak across the grid condenser brings the tube back to a condition where it can receive the next signal. While the sending key is depressed the current in the receiver is decreasing and increasing once for every signal which is being transmitted. For each decrease and increase in current the diaphragm of the telephone receiver makes one vibration. What the listener then hears is a musical note with a frequency corresponding 173to that number of vibrations a second, that is, a note with a frequency of one thousand cycles per second. He hears a note of frequency about that of two octaves above middle *C* on the piano. There are usually other notes present at the same time and the sound is not like that of any musical instrument.

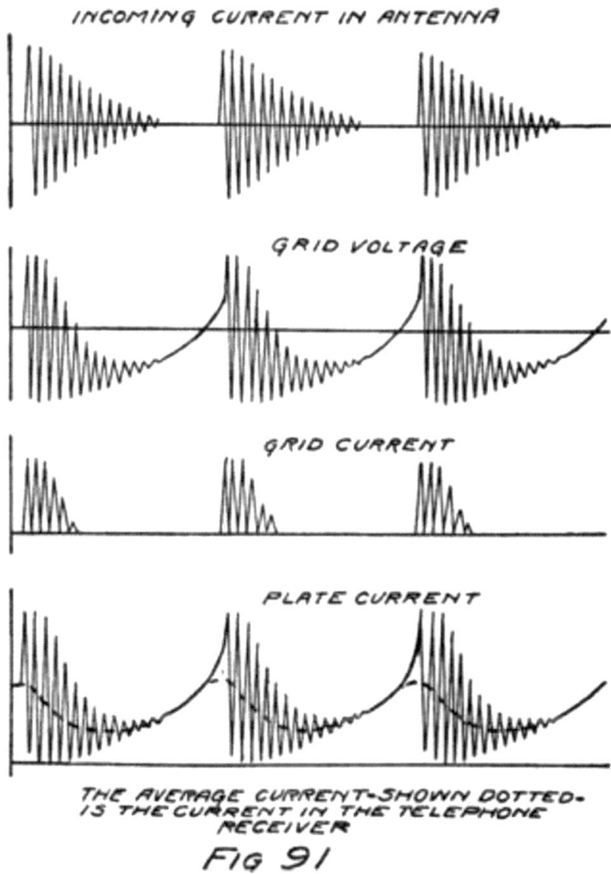

FIG 91

If the key is held down a long time for a dash the listener hears this note for a corresponding time. If it is depressed only about a third of that time 174so as to send a dot, the listener hears the note for a shorter time and interprets it to mean a dot.

In Fig. 91 I have drawn a sketch to show the e. m. f. which the signals from a spark set impress on the grid of a detector and to show how the plate current varies if there is a condenser and leak in the grid circuit. I have only shown three signals in succession. If the operator sends at the rate of about twenty words a minute a dot is formed by about sixty of these signals in succession.

The frequency of the alternations in one of the little signals will depend upon the wave length which the sending operator is using. If he uses the wave length of 600 meters, as ship stations do, he will send with a radio frequency of 500,000 cycles a second. Since the signals are at the rate of a thousand a second each one is made up of 500 complete cycles of the current in the antenna. It would be impracticable therefore to show you a complete picture of the signal from a spark set. I have, however, lettered the figure quite completely to cover what I have just told you.

If the grid-condenser and its leak are so chosen as to work well for signals from a 500-cycle spark set they will also work well for the notes in human speech which are about 1000 cycles a second in frequency. The detecting circuit will not, however, work so well for the other notes which are in the human voice and are necessary to speech. For example, if notes of about 2000 cycles a second are involved in the speech which is being transmitted, 175the leak across the condenser will not work fast enough. On the other hand, for the very lowest notes in the voice the leak will work too fast and such variations in the signal current will not be detected as efficiently as are those of 1000 cycles a second.

You can see that there is always a little favoritism on the part of the grid-condenser detector. It doesn't exactly reproduce the variations in intensity of the radio signal which were made at the sending station. It distorts a little. As amateurs we usually forgive it that distortion because it is so efficient. It makes so large a change in the current through the telephone when it receives a signal that we can use it to receive much weaker signals, that is, signals from smaller or more distant sending stations, than we can receive with the arrangement described in Letter 14.

176 LETTER 18
AMPLIFIERS AND THE REGENERATIVE CIRCUIT

My Dear Receiver:

There is one way of making an audion even more efficient as a detector than the method described in the last letter. And that is to make it talk to itself.

Suppose we arrange a receiving circuit as in Fig. 92. It is exactly like that of Fig. 90 of the previous letter except for the fact that the current in the plate circuit passes through a little coil, L_T, which is placed near the coil L and so can induce in it an e. m. f. which will correspond in intensity and wave form to the current in the plate circuit.

If we should take out the grid condenser and its leak this circuit would be like that of Fig. 54 in Letter 13 which we used for a generator of high-frequency alternating currents. You remember how that circuit operates. A small effect in the grid circuit produces a large effect in the plate circuit. Because the plate circuit is coupled to the grid circuit the grid is again affected and so there is a still larger effect in the plate circuit. And so on, until the current in the plate circuit is swinging from zero to its maximum possible value.

What happens depends upon how closely the coils L and L_T are coupled, that is, upon how much the 177 changing current in one can affect the other. If they are turned at right angles to each other, so that there is no possible mutual effect we say there is "zero coupling."

Start with the coils at right angles to each other and turn L_T so as to bring its windings more and more parallel to those of L. If we want L_T to have a large effect on L its windings should be parallel and also in the same direction just as they were in Fig. 54 of Letter 13 to which we just referred. As we approach nearer to that position the current in L_T induces more and more e. m. f. in coil L. For some position of the two coils, and the actual position depends on the tube we are using, there will be enough effect from the plate circuit upon the grid circuit so that there will be continuous oscillations.

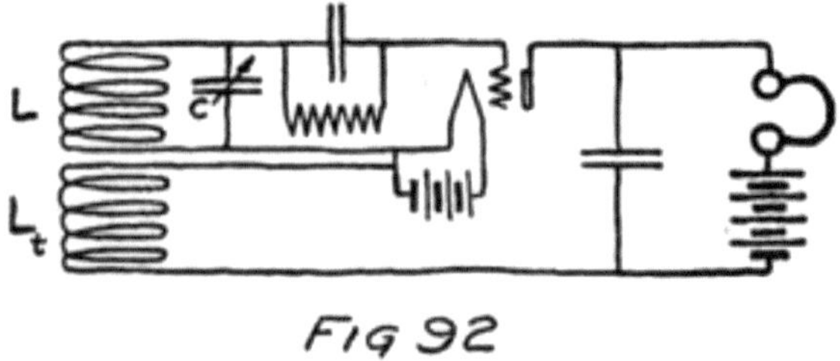

FIG 92

We want to stop just short of this position. There will then be no continuous oscillations; but if any changes do take place in the plate current they will affect the grid. And these changes in the grid voltage will result in still larger changes in the plate current.

Now suppose that there is coming into the detector circuit of Fig. 92 a radio signal with, speech significance. 178The current in the plate circuit varies accordingly. So does the current in coil L_T which is in the plate circuit. But this current induces an e. m. f. in coil L and this adds to the e. m. f. of the incoming signal so as to make a greater variation in the plate current. This goes on as long as there is an incoming signal. Because the plate circuit is coupled to the grid circuit the result is a larger e. m. f. in the grid circuit than the incoming signal could set up all by itself.

You see now why I said the tube talked to itself. It repeats to itself whatever it receives. It has a greater strength of signal to detect than if it didn't repeat. Of course, it detects also just as I told you in the preceding letter.

In adjusting the coupling of the two coils of Fig. 92 we stopped short of allowing the tube circuit to oscillate and to generate a high frequency. If we had gone on increasing the coupling we should have reached a position where steady oscillations would begin. Usually this is marked by a little click in the receiver. The reason is that when the tube oscillates the average current in the plate circuit is not the same as the steady current which ordinarily flows between filament and plate. There is a sudden change, therefore, in the average current in the plate circuit when the tube starts to oscillate. You remember that what affects the receiver is the average current in the plate circuit. So the receiver diaphragm suddenly changes position as the tube starts to oscillate and a listener hears a little click.

179The frequency of the alternating current which the tube produces depends upon the tuned circuit formed by L and C. Suppose that this frequency is not the same as that to which the receiving antenna is tuned. What will happen?

There will be impressed on the grid of the tube two alternating e. m. f.'s, one due to the tube's own oscillations and the other incoming from the distant transmitting station. The two e. m. f. 's are both active at once so that at each instant the e. m. f. of the grid is really the sum of these two e. m. f.'s. Suppose at some instant both e. m. f.'s are acting to make the grid positive. A little later one of them will be trying to make the grid negative while the other is still trying to make it positive. And later still when the first e. m. f. is ready again to make the grid positive the second will be trying to make it negative.

It's like two men walking along together but with different lengths of step. Even if they start together with their left feet they are soon so completely out of step that one is putting down his right foot while the other is putting down his left. A little later, but just for an instant, they are in step again. And so it goes. They are in step for a moment and then completely out of step. Suppose one of them makes ten steps in the time that the other makes nine. In that time they will be once in step and once completely out of step. If one makes ten steps while the other does eight this will happen twice.

The same thing happens in the audion detector 180circuit when two e. m. f.'s which differ slightly in frequency are simultaneously impressed on the grid. If one e. m. f. passes through ten complete cycles while the other is making eight cycles, then during that time they will twice be exactly in step, that is, "in phase" as we say. Twice in that time they will be exactly out of step, that is, exactly "opposite in phase." Twice in that time the two e. m. f.'s will aid each other in their effects on the grid and twice they will exactly oppose. Unless they are equal in amplitude there will still be a net e. m. f. even when they are exactly opposed. The result of all this is that the average current in the plate circuit of the detector will alternately increase and decrease twice during this time.

The listener will then hear a note of a frequency equal to the difference between the frequencies of the two e. m. f.'s which are being simultaneously impressed on the grid of the detector. Suppose the incoming signal has a frequency of 100,000 cycles a second but that the detector tube is oscillating in its own circuit at the rate of 99,000 cycles per second, then the listener will hear a note of 1000 cycles per second. One thousand times each second the two e. m. f.'s will be exactly in phase and one thousand times each second they will be exactly opposite in phase. The voltage applied to the grid will be a maximum one thousand times a second and alternately a minimum. We can think of it, then, as if there were impressed on the grid of the detector a high-frequency signal which varied in intensity one thousand times a 181second. This we know will produce a corresponding variation in the current through the telephone receiver and thus give rise to a musical note of about two octaves above middle *C* on the piano.

This circuit of Fig. 92 will let us detect signals which are not varying in intensity. And consequently this is the method which we use to detect the telegraph signals which are sent out by such a "continuous wave transmitter" as I showed you at the end of Letter 13.

When the key of a C-W transmitter is depressed there is set up in the distant receiving-antenna an alternating current. This current doesn't vary in strength. It is there as long as the sender has his key down. Because, however, of the effect which I described above there will be an audible note from the telephone receiver if the detector tube is oscillating at a frequency within two or three thousand cycles of that of the transmitting station.

This method of receiving continuous wave signals is called the "heterodyne" method. The name comes from two Greek words, "dyne" meaning "force" and the other part meaning "different." We receive by combining two different electron-moving-forces, one produced by the distant sending-station and the other produced locally at the receiving station. Neither by itself will produce any sound, except a click when it starts. Both together produce a musical sound in the telephone receiver; and the frequency of that note is the difference of the two frequencies.

182There are a number of words used to describe this circuit with some of which you should be familiar. It is sometimes called a "feed-back" circuit because part of the output of the audion is fed back into its input side. More generally it is known as the "regenerative circuit" because the tube keeps on generating an alternating current. The little coil which is used to feed back into the grid circuit some of the effects from the plate circuit is sometimes called a "tickler" coil.

It is not necessary to use a grid condenser in a feed-back circuit but it is perhaps the usual method of detecting where the regenerative circuit is used. The whole value of the regenerative circuit so far as receiving is concerned is in the high efficiency which it permits. One tube can do the work of two.

We can get just as loud signals by using another tube instead of making one do all the work. In the regenerative circuit the tube is performing two jobs at once. It is detecting but it is also amplifying. [9] By "amplifying" we mean making an e. m. f. larger than it is without changing the shape of its picture, that is without changing its "wave form."

To show just what we mean by amplifying we must look again at the audion and see how it acts. You know that a change in the grid potential makes a change in the plate current. Let us arrange an audion in a circuit which will tell us a little more of what happens. Fig. 93 shows the circuit.

183This circuit is the same as we used to find the audion characteristic except that there is a clip for varying the number of batteries in the plate circuit and a voltmeter for measuring their e. m. f. We start with the grid at zero potential and the usual number of batteries in the plate circuit. The voltmeter tells us the e. m. f. We read the ammeter in the plate circuit and note what that current is. Then we shift the slider in the grid circuit so as to give the grid a small potential. The current in the plate circuit changes. We can now move the clip on the B-batteries so as to bring the current in this circuit back to its original value. Of course, if we make the grid positive we move the clip so as to use fewer cells of the B-battery. On the other hand if we make the grid negative we shall need more e. m. f. in the plate circuit. In either case we shall find that we need to make a very much larger change in the voltage of the plate circuit than we have made in the voltage of the grid circuit.

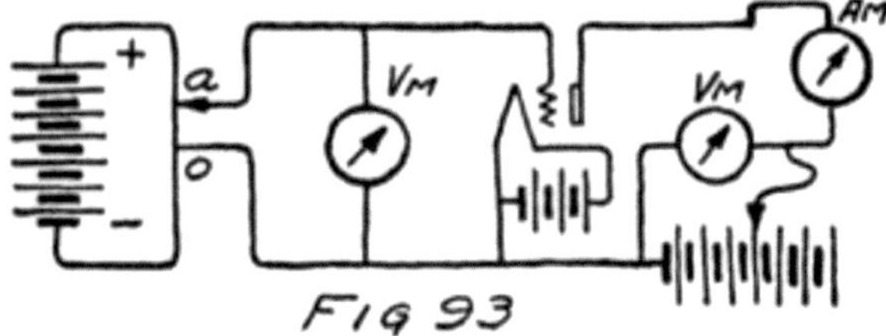

FIG 93

Usually we perform the experiment a little differently so as to get more accurate results. We read the voltmeter in the plate circuit and the ammeter in that circuit. Then we change the number of batteries which we are using in the plate circuit. That changes 184the plate current. The next step is to shift the slider in the grid circuit until we have again the original value of current in the plate circuit. Suppose that the tube is ordinarily run with a plate voltage of 40 volts and we start with that e. m. f. on the plate. Suppose that we now make it 50 volts and then vary the position of the slider in the grid circuit until the ammeter reads as it did at the start. Next we read the voltage impressed on the grid by reading the voltmeter in the grid circuit. Suppose it reads 2 volts. What does that mean?

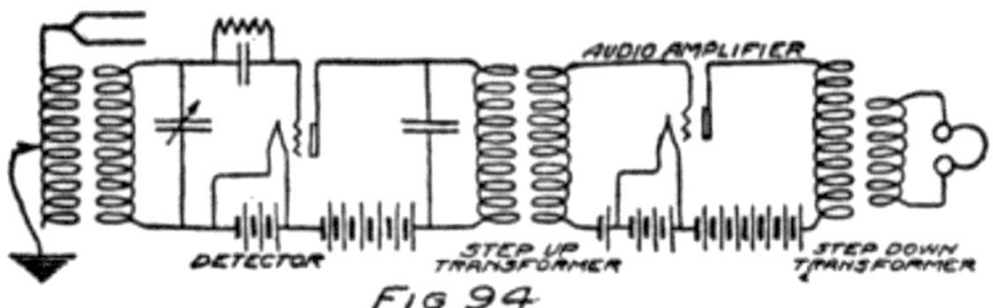

FIG 94

It means that two volts in the grid circuit have the same effect on the plate current as ten volts in the plate circuit. If we apply a volt to the grid circuit we get five times as large an effect in the plate circuit as we would if the volt were applied there. We get a greater effect, the effect of more volts, by applying our voltage to the grid. We say that the tube acts as an "amplifier of voltage" because we can get a larger effect than the number of volts which we apply would ordinarily entitle us to.

Now let's take a simple case of the use of an audion as an amplifier. Suppose we have a receiving circuit with which we find that the signals are 185not easily understood because they are too weak. Let this be the receiving circuit of Fig. 88 which I am reproducing here as part of Fig. 94.

We have replaced the telephone receiver by a "transformer." A transformer is two coils, or windings, coupled together. An alternating current in one will give rise to an alternating current in the other. You are already familiar with the idea but this is our first use of the word. Usually we call the first coil, that is the one through which the alternating current flows, the "primary" and the second coil, in which a current is induced, the "secondary."

The secondary of this transformer is connected to the grid circuit of another vacuum tube, to the plate circuit of which is connected another transformer and the telephone receiver. The result is a detector and "one stage of amplification."

The primary of the first transformer, so we shall suppose, has in it the same current as would have been in the telephone. This alternating current induces in the secondary an e. m. f. which has the same variations as this current. This e. m. f. acts on the grid of the second tube, that is on the amplifier. Because the audion amplifies, the e. m. f. acting on the telephone receiver is larger than it would have been without the use of this audion. And hence there is a greater response on the part of its diaphragm and a louder sound.

In setting up such a circuit as this there are several things to watch. For some of these you will 186have to rely on the dealer from whom you buy your supplies and for the others upon yourself. But it will take another letter to tell you of the proper precautions in using an audion as an amplifier.

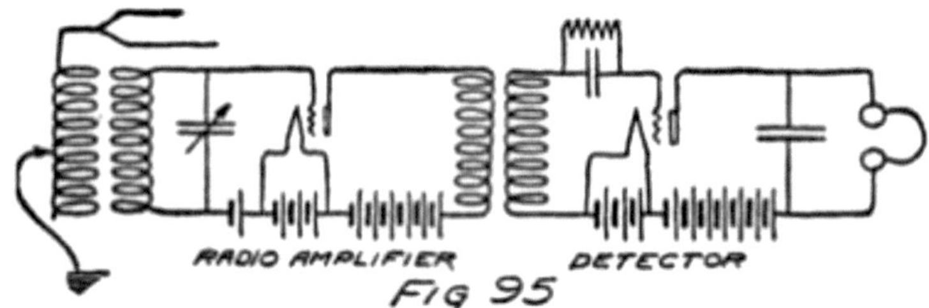

In the circuit which I have just described an audion is used to amplify the current which comes from the detector before it reaches the telephone receiver. Sometimes we use an audion to amplify the e. m. f. of the signal before impressing it upon the grid of the detector. Fig. 95 shows a circuit for doing that. In the case of Fig. 94 we are amplifying the audio-frequency current. In that of Fig. 95 it is the radio-frequency effect which is amplified. The feed-back or regenerative circuit of Fig. 92 is a one-tube circuit for doing the same thing as is done with two tubes in Fig 95.

[9]

There is always some amplification taking place in an audion detector but the regenerative circuit amplifies over and over again until the signal is as large as the tube can detect.

 LETTER 19
THE AUDION AMPLIFIER AND ITS CONNEC-TIONS

In our use of the audion we form three circuits. The first or A-circuit includes the filament. The B-circuit includes the part of the tube between filament and plate. The C-circuit includes the part between filament and grid. We sometimes speak of the C-circuit as the "input" circuit and the B-circuit as the "output" circuit of the tube. This is because we can put into the grid-filament terminals an e. m. f. and obtain from the plate-filament circuit an effect in the form of a change of current.

FIG 96

Suppose we had concealed in a box the audion and circuit of Fig. 96 and that only the terminals which are shown came through the box. We are given a battery and an ammeter and asked to find out all we can as to what is between the terminals *F* and *G*. We connect the battery and ammeter in series with these terminals. No current flows through the circuit. We reverse the battery but no current flows in the opposite direction. Then we reason that there is an open-circuit between *F* and *G*.

As long as we do not use a higher voltage than 188that of the C-battery which is in the box no current can flow. Even if we do use a higher voltage than the "negative C-battery" of the hidden grid-circuit there will be a current only when the external battery is connected so as to make the grid positive with respect to the filament.

Now suppose we take several cells of battery and try in the same way to find what is hidden between the terminals *P* and *F*. We start with one battery and the ammeter as before and find that if this battery is connected so as to make *P* positive with respect to *F*, there is a feeble current. We increase the battery and find that the current is increased. Two cells, however, do not give exactly twice the current that one cell does, nor do three give three times as much. The current does not increase proportionately to the applied voltage. Therefore we reason that whatever is between *P* and *F* acts like a resistance but not like a wire resistance.

Then, we try another experiment with this hidden audion. We connect a battery to *G* and *F*, and note what effect it has on the current which our other battery is sending through the box between *P* and *F*. There is a change of current in this circuit, just as if our act of connecting a battery to *G-F* had resulted in connecting a battery in series with the *P-F* circuit. The effect is exactly as if there is inside the box a battery which is connected into the hidden part of the circuit *P-F*. This concealed battery, which now starts to act, appears to be several times stronger than the battery which is connected to *G-F*.

189Sometimes this hidden battery helps the B-battery which is on the outside; and sometimes it seems to oppose, for the current in the *P-F* circuit either increases or decreases, depending upon how we connect the battery to *G* and *F*. The hidden battery is always larger than our battery connected to *G* and *F*. If we arrange rapidly to reverse the battery connected to *G-F* it appears as if there is inside the box in the *P-F* circuit an alternator, that is, something which can produce an alternating e. m. f.

All this, of course, is merely a review statement of what we already know. These experiments are interesting, however, because they follow somewhat those which were performed in studying the audion and finding out how to make it do all the wonderful things which it now can.

As far as we have carried our series of experiments the box might contain two separate circuits. One between *G* and *F* appears to be an open circuit. The other appears to have in it a resistance and a battery (or else an alternator). The e. m. f. of the battery, or alternator, as the case may be, depends on what source of e. m. f. is connected to *G-F*. Whatever that e. m. f. is, there is a corresponding kind of e. m. f. inside the box but one several times larger.

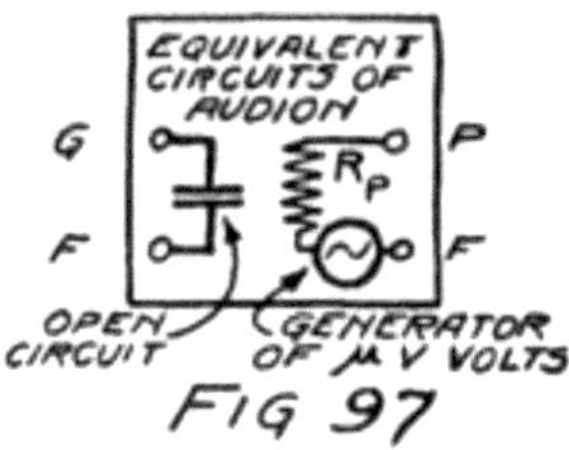

We might, therefore, pay no further attention to what is actually inside the box or how all these effects are brought about. We might treat the entire box 190as if it was formed by two separate circuits as shown in Fig. 97. If we do so, we are replacing the box by something which is equivalent so far as effects are concerned, that is we are replacing an actual audion by two circuits which together are equivalent to it.

The men who first performed such experiments wanted some convenient way of saying that if an alternator, which has an e. m. f. of V volts, is connected to F and G, the effect is the same as if a much stronger alternator is connected between F and P. How much stronger this imaginary alternator is depends upon the design of the audion. For some audions it might be five times as strong, for other designs 6.5 or almost any other number, although usually a number of times less than 40. They used a little Greek letter called "mu" to stand for this number which depends on the design of the tube. Then they said that the hidden alternator in the output circuit was mu times as strong as the actual alternator which was applied between the grid and the filament. Of course, instead of writing the sound and name of the letter they used the letter μ itself. And that is what I have done in the sketch of Fig. 97.

Now we are ready to talk about the audion as an amplifier. The first thing to notice is the fact that we have an open circuit between F and G. This is true as long as we don't apply an e. m. f. large enough to overcome the C-battery of Fig. 96 and thus let the grid become positive and attract electrons from the filament. We need then spend no further time thinking 191 about what will happen in the circuit G-F, for there will be no current.

As to the circuit F-P, we can treat it as a resistance in series with which there is a generator μ times as strong as that which is connected to F and G. The next problem is how to get the most out of this hidden generator. We call the resistance which the tube offers to the passage of electrons between P and F the "internal resistance" of the plate circuit of the tube. How large it is depends upon the design of tube. In some tubes it may be five or six thousand ohms, and in others several times as high. In the large tubes used in high-powered transmitting sets it is much less. Since it will be different in different cases we shall use a symbol for it and say that it is R_P ohms.

Then one rule for using an audion as an amplifier is this: To get the most out of an audion see that the telephone, or whatever circuit or piece of apparatus is connected to the output terminals, shall have a resistance of R_P ohms. When the resistance of the circuit, which an audion is supplying with current, is the same as the internal resistance of the output side of the tube, then the audion gives its greatest output. That is the condition for the greatest "amount of energy each second," or the "greatest power" as we say.

That rule is why we always select the telephone receivers which we use with an audion and always ask carefully as to their resistance when we buy. Sometimes, however, it is not practicable to use receivers 192of just the right resistance. Where we connect the output side of an audion to some other circuit, as where we let one audion supply another, it is usually impossible to follow this rule without adding some special apparatus.

This leads to the next rule: If the telephone receiver, or the circuit, which we wish to connect to the output of an audion, does not have quite nearly a resistance of R_P ohms we use a properly designed transformer as we have already done in Figs. 94 and 95.

A transformer is two separate coils coupled together so that an alternating current in the primary will induce an alternating current in the secondary. Of course, if the secondary is open-circuited then no current can flow but there will be induced in it an e. m. f. which is ready to act if the circuit is closed. Transformers have an interesting ability to make a large resistance look small or vice versa. To show you why, I shall have to develop some rules for transformers.

Suppose you have an alternating e. m. f. of ten volts applied to the primary of an iron-cored transformer which has ten turns. There is one volt applied to each turn. Now, suppose the secondary has only one turn. That one turn has induced in it an alternating e. m. f. of one volt. If there are more turns of wire forming the secondary, then each turn has induced in it just one volt. But the e. m. f.'s of all these turns add together. If the secondary has twenty turns, there is induced in it a total of twenty 193volts. So the first rule is this: In a transformer the number of volts in each turn of wire is just the same in the secondary as in the primary.

If we want a high-voltage alternating e. m. f. all we have to do is to send an alternating current through the primary of a transformer which has in the secondary, many times more turns of wire than it has in the primary. From the secondary we obtain a higher voltage than we impress on the primary.

You can see one application of this rule at once. When we use an audion as an amplifier of an alternating current we send the current which is to be amplified through the primary of a transformer, as in Fig. 94. We use a transformer with many times more turns on the secondary than on the primary so as to apply a large e. m. f. to the grid of the amplifying tube. That will mean a large effect in the plate circuit of the amplifier.

You remember that the grid circuit of an audion with a proper value of negative C-battery is really open-circuited and no current will flow in it. For that case we get a real gain by using a "step-up" transformer, that is, one with more turns in the secondary than in the primary.

It looks at first as if a transformer would always give a gain. *If we mean a gain in energy it will not* although we may use it, as we shall see in a minute, to permit a vacuum tube to work into an output circuit more efficiently than it could without the transformer. We cannot have any more energy 194 in the secondary circuit of a transformer than we give to the primary.

Suppose we have a transformer with twice as many turns on the secondary as on the primary. To the primary we apply an alternating e. m. f. of a certain number of volts. In the secondary there will be twice as many volts because it has twice as many turns. The current in the secondary, however, will be only half as large as is the current in the primary. We have twice the force in the secondary but only half the electron stream.

It is something like this: You are out coasting and two youngsters ask you to pull them and their sleds up hill. You pull one of them all the way and do a certain amount of work. On the other hand suppose you pull them both at once but only half way up. You pull twice as hard but only half as far and you do the same amount of work as before.

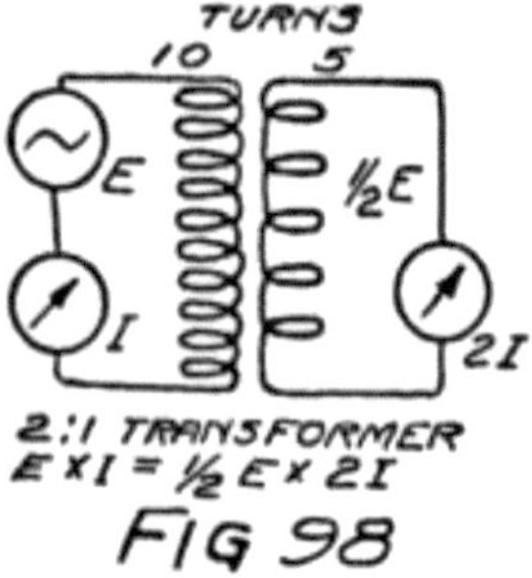

We can't get more work out of the secondary of a transformer than we do in the primary. If we design the transformer so that there is a greater pull (e. m. f.) in the secondary the electron stream in the secondary will be correspondingly smaller.

You remember how we measure resistance. We divide the e. m. f. (number of volts) by the current (number of amperes) to find the resistance (number of ohms). Suppose we do that for the primary and for the 195secondary of the transformer of Fig. 98 which we are discussing. See what happens in the secondary. There is only half as much voltage but twice as much current. It looks as though the secondary had one-fourth as much resistance as the primary. And so it has, but we usually call it "impedance" instead of resistance because straight wires resist but coils or condensers impede alternating e. m. f.'s.

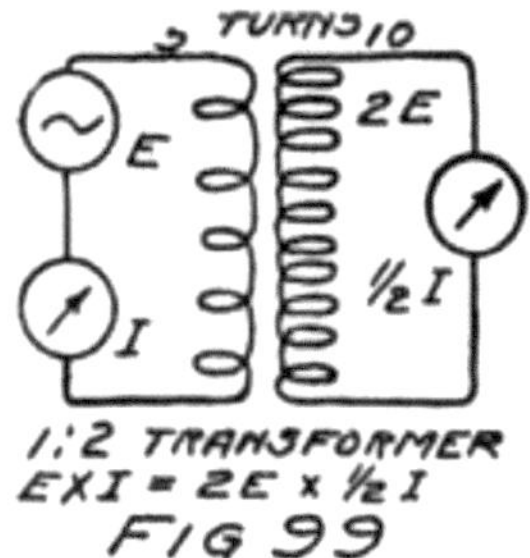

Before we return to the question of using a transformer in an audion circuit let us turn this transformer around as in Fig. 99 and send the current through the side with the larger number of windings. Let's talk of "primary" and "secondary" just as before but, of course, remember that now the primary has twice the turns of the secondary. On the secondary side we shall have only half the current, but there will be twice the e. m. f. The resistance of the secondary then is four times that of the primary.

Now return to the amplifier of Fig. 94 and see what sort of a transformer should be between the plate circuit of the tube and the telephone receivers. Suppose the internal resistance of the tube is 12,000 ohms and the resistance of the telephones is 3,000 ohms. Suppose also that the resistance (really impedance) of the primary side of the transformer which we just considered is 12,000 ohms. The impedance of its secondary will be a quarter of this or 3,000 ohms. 196If we connect such a transformer in the circuit, as shown, we shall obtain the greatest output from the tube.

In the first place the primary of the transformer has a number of ohms just equal to the internal resistance of the tube. The tube, therefore, will give its best to that transformer. In the second place the secondary of the transformer has a resistance just equal to the telephone receivers so it can give its best to them. The effect of the transformer is to make the telephones act as if they had four times as much resistance and so were exactly suited to be connected to the audion.

This whole matter of the proper use of transformers is quite simple but very important in setting up vacuum-tube circuits. To overlook it in building or buying your radio set will mean poor efficiency. Whenever you have two parts of a vacuum-tube circuit to connect together be sure and buy only a transformer which is designed to work between the two impedances (or resistances) which you wish to connect together.

There is one more precaution in connection with the purchase of transformers. They should do the same thing for all the important frequencies which they are to transmit. If they do not, the speech or signals will be distorted and may be unintelligible.

If you take the precautions which I have mentioned your radio receiving set formed by a detector and one amplifier will look like that of Fig. 94. That is only one possible scheme of connections. You can use 197any detector circuit which you wish, [10] one with a grid condenser and leak, or one arranged for feed-back In either case your amplifier may well be as shown in the figure.

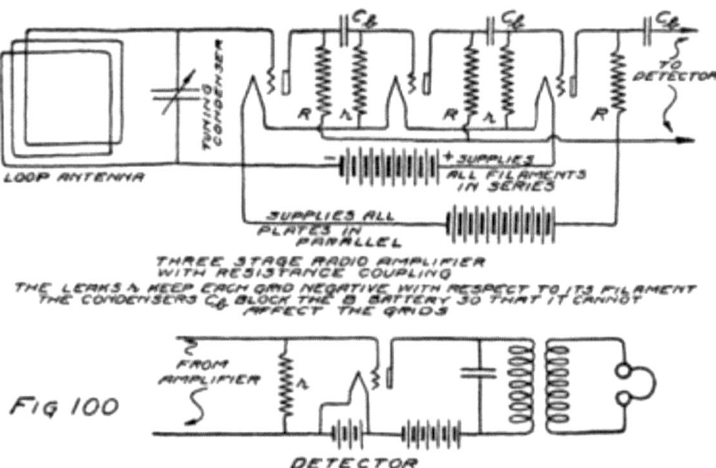

FIG 100

198The circuit I have described uses an audion to amplify the audio-frequency currents which come from the detector and are capable of operating the telephones. In some cases it is desirable to amplify the radio signals before applying them to the detector. This is especially true where a "loop antenna" is being used. Loop antennas are smaller and more convenient than aërials and they also have certain abilities to select the signals which they are to receive because they receive best from stations which lie along a line drawn parallel to their turns. Unfortunately, however, they are much less efficient and so require the use of amplifiers.

With a small loop made by ten turns of wire separated by about a quarter of an inch and wound on a square mounting, about three feet on a side, you will usually require two amplifiers. One of these might be used to amplify the radio signals before detection and the other to amplify after detection. To tune the loop for broadcasts a condenser of about 0.0005 mf. will be needed. The diagram of Fig. 100 shows the complete circuit of a set with three stages of radio-amplification and none of audio.

[10]

Except for patented circuits. See p. 224.

199 LETTER 20
TELEPHONE RECEIVERS AND OTHER ELEC-
TROMAGNETIC DEVICES

In an earlier letter when we first introduced a telephone receiver into a circuit I told you something of how it operates. I want now to tell why and also of some other important devices which operate for the same reason.

You remember that a stream of electrons which is starting or stopping can induce the electrons of a neighboring parallel circuit to start off in parallel paths. We do not know the explanation of this. Nor do we know the explanation of another fact which seems to be related to this fact of induction and is the basis for our explanations of magnetism.

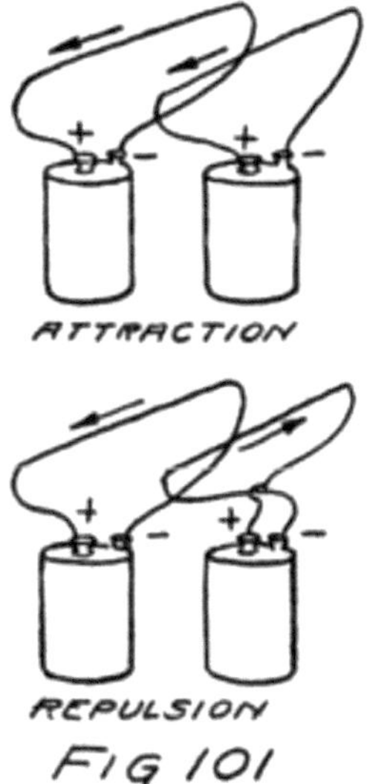

If two parallel wires are carrying steady electron streams in the same general direction the wires attract each other. If the streams are oppositely directed the wires repel each other. Fig. 101 illustrates 200this fact. If the streams are not at all in the same direction, that is, if they are at right angles, they have no effect on each other.

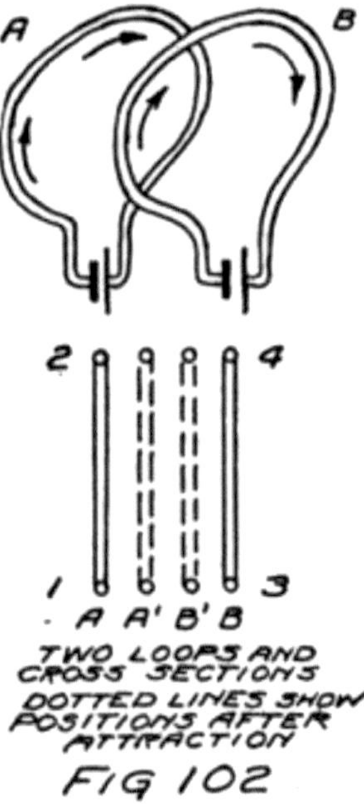

These facts, of the attraction of electron streams which are in the same direction and repulsion of streams in opposite directions, are all that one need remember to figure out for himself what will happen under various conditions. For example, if two coils of wire are carrying currents what will happen is easily seen. Fig. 102 shows the two coils and a section through them.

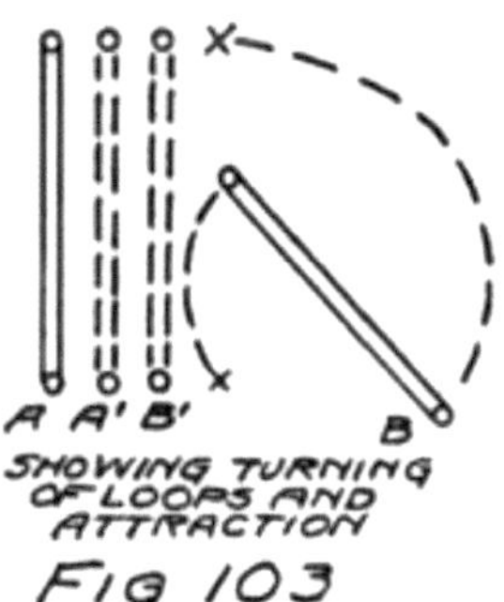

Looking at this cross section we seem to have four wires, *1* and *2* of coil *A* and *3* and *4* of coil *B*. You see at once that if the coils are free to move they will move into the dotted positions shown in Fig 102, because wire *1* attracts wire *3* and repels wire *4*, while wire *2* attracts wire *4* and repels wire *3*. If necessary, and if they are free to move, the coils will turn completely around to get to this position. I have shown such a case in Fig. 103.

201Wires which are not carrying currents do not behave in this way. The action is due, but how we don't yet know, to the motions of the electrons. As far as we can explain it to-day, the attraction of two wires which are carrying currents is due to the attraction of the two streams of electrons. Of course these electrons are part of the wires. They can't get far away from the stay-at-home electrons and the nuclei of the atoms which form the wires. In fact it is these nuclei which keep the wandering electrons within the wires. The result is that if the streams of electrons are to move toward each other the wires must go along with them.

If the wires are held firmly the electron streams cannot approach one another for they must stay in the wires. Wires, therefore, perform the important service of acting as paths for electrons which are traveling as electric currents. There are other ways in which electrons can be kept in a path, and other means beside batteries for keeping them going. It doesn't make any difference so far as the attraction or the repulsion is concerned why they are following a certain path or why they stay in it. So far as we know two streams of electrons, following parallel paths, will always, behave just like the two streams of Fig. 101.

Suppose, for example, there were two atoms which were each formed by a nucleus and a number of electrons swinging around about the nucleus as pictured in 202Fig. 104. The electrons are going of their own accord and the nucleus keeps them from flying off at a tangent, the way mud flies from the wheel of an automobile. Suppose these two atoms are free to turn but not to move far from their present positions. They will turn so as to make their electron paths parallel just as did the loops of Fig. 102.

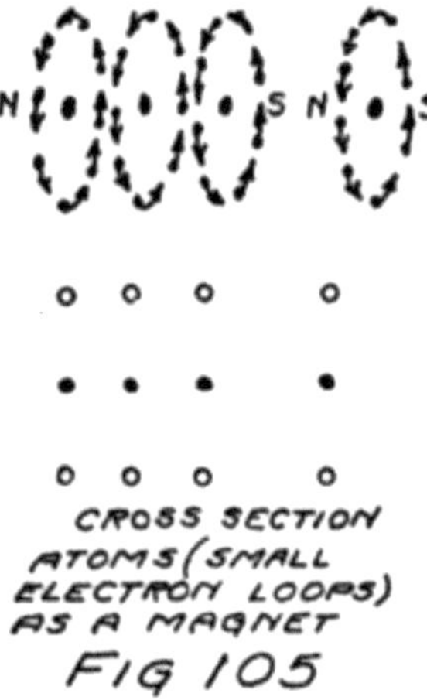

Now, I don't say that there are any atoms at all like the ones I have pictured. There is still a great deal to be learned about how electrons act inside different kinds of atoms. We do know, however, that the atoms of iron act just as if they were tiny loops with electron streams.

190

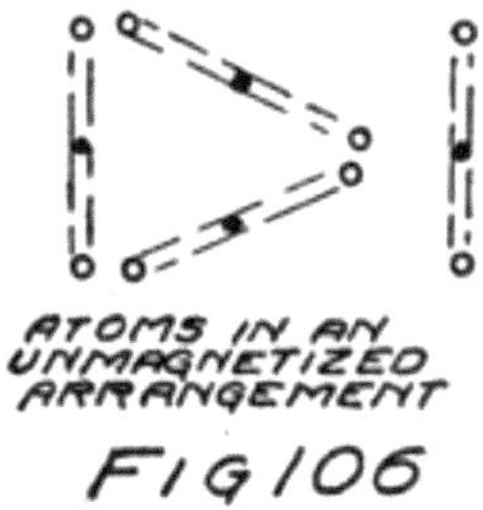

203Suppose we had several loops and that they were lined up like the three loops in Fig. 105. You can see that they would all attract the other loop, on the right in the figure. On the other hand if they were grouped in the triangle of Fig. 106 they would barely affect the loop because they would be pulling at cross purposes. If a lot of the tiny loops of the iron atoms are lined up so as to act together and attract other loops, as in the first figure, we say the iron is magnetized and is a magnet. In an ordinary piece of iron, however, the atoms are so grouped that they don't pull together but like the loops of our second figure pull in different directions and neutralize each other's efforts so that there is no net effect.

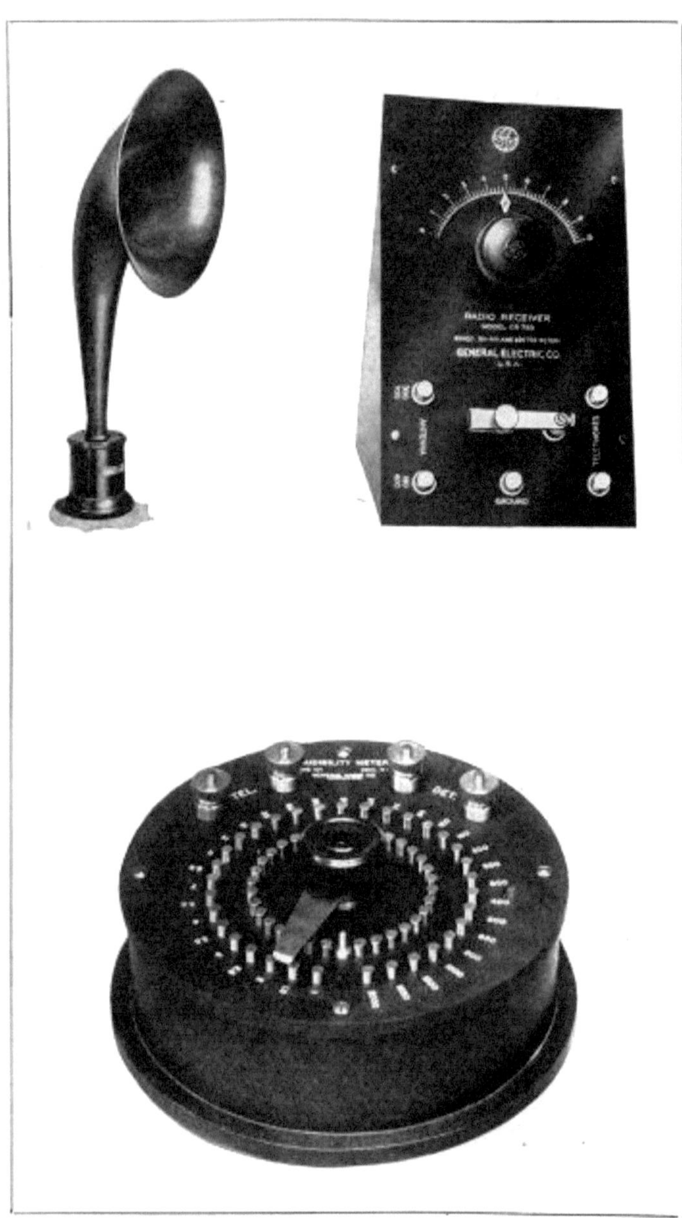

204And like the loops of Fig. 106 the atoms in an unmagnetized piece of iron are pretty well satisfied to stay as they are without all lining up to pull together. To magnetize the iron we must force some of these atomic loops to turn part way around. That can be done by bringing near them a strong magnet or a coil of wire which is carrying a current. Then the atoms are forced to turn and if enough turn so that there is an appreciable effect then the iron is magnetized. The more that are properly turned the stronger is the magnet. One end or "pole" we call north-seeking and the other south-seeking, because a magnetized bar of iron acts like a compass needle.

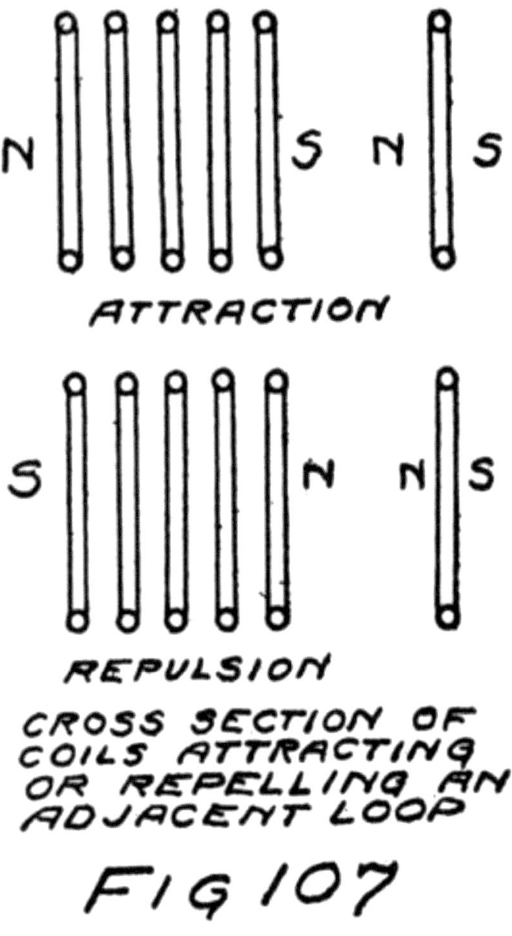

A coil of wire, carrying a current, acts just like a magnet because its larger loops are all ready to pull together. I have marked the coil of Fig. 107 with *N* and *S* for north and south. If the electron stream in it is reversed the "polarity" is reversed. There is a simple rule for this. Partially close your left hand so that the fingers form loops. Let the thumb stick out at right angles to these loops. If the 205 electron streams are flowing around the loops of a coil in the same direction as your fingers point then your thumb is the *N* pole and the coil will repel the north poles of other loops or magnets in the direction in

194

which your thumb points. If you know the polarity already there is a simple rule for the repulsion or attraction. Like poles repel, unlike poles attract.

From what has been said about magnetism you can now understand why in a telephone receiver the current in the winding can make the magnet stronger. It does so because it makes more of the atomic loops of the iron turn around and help pull. On the other hand if the current in the winding is reversed it will turn some of the loops which are already helping into other positions where they don't help and may hinder. If the current in the coil is to help, the electron stream in it must be so directed that the north pole of the coil is at the same end as the north pole of the magnet.

This idea of the attraction or repulsion of electron streams, whether in coils of wire or in atoms of iron and other magnetizable substances, is the fundamental idea of most forms of telephone receivers, of electric motors, and of a lot of other devices which we call "electromagnetic."

The ammeters and voltmeters which we use for the measurement of audion characteristics and the like are usually electromagnetic instruments. Ammeters and voltmeters are alike in their design. Both are sensitive current-measuring instruments. In the case of the voltmeter, as you know, we have 206a large resistance in series with the current-measuring part for the reason of which I told in Letter 8. In the case of ammeters we sometimes let all the current go through the current-measuring part but generally we let only a certain fraction of it do so. To pass the rest of the current we connect a small resistance in parallel with the measuring part. In both types of instruments the resistances are sometimes hidden away under the cover. Both instruments must, of course, be calibrated as I have explained before.

In the electromagnetic instruments there are several ways of making the current-measuring part. The simplest is to let the current, or part of it, flow through a coil which is pivoted between the N and S poles of a strong permanent magnet. A spring keeps the coil in its zero position and if the current makes the coil turn it must do so against this spring. The stronger the current in the coil the greater the interaction of the loops of the coil and those of the iron atoms and hence the further the coil will turn. A pointer attached to the coil indicates how far; and the number of volts or amperes is read off from the calibrated scale.

Such instruments measure direct-currents, that is, steady streams of electrons in one direction. To measure an alternating current or voltage we can use a hot-wire instrument or one of several different types of electromagnetic instruments. Perhaps the simplest of these is the so-called "plunger type." The alternating current flows in a coil; and a piece of 207soft iron is so pivoted that it can be attracted and moved into the coil. Soft iron does not make a good permanent magnet. If you put a piece of it inside a coil which is carrying a steady current it becomes a magnet but about as soon as you interrupt the current the atomic loops of the iron stop pulling together. Almost immediately they turn into all sorts of positions and form little self-satisfied groups which don't take any interest in the outside world. (That isn't true of steel, where the atomic loops are harder to turn and to line up, but are much more likely to stay in their new positions.)

Because the plunger in an alternating-current ammeter is soft iron its loops line up with those of the coil no matter which way the electron stream happens to be going in the coil. The atomic magnets in the iron turn around each time the current reverses and they are always, therefore, lined up so that the plunger is attracted. If the plunger has much inertia or if the oscillations of the current are reasonably frequent the plunger will not move back and forth with each reversal of the current but will take an average position. The stronger the a-c (alternating current) the farther inside the coil will be this position of the plunger. The position of the plunger becomes then a measure of the strength of the alternating current.

Instruments for measuring alternating e. m. f.'s and currents, read in volts and in amperes. So far I haven't stopped to tell what we mean by one ampere of alternating current. You know from Letter 7 208what we mean by an ampere of d-c (direct current). It wasn't necessary to explain before because I told you only of hot-wire instruments and they will read the same for either d-c or a-c.

When there is an alternating current in a wire the electrons start, rush ahead, stop, rush back, stop, and do it all over again and again.

That heats the wire in which it happens. If an alternating stream of electrons, which are doing this sort of thing, heats a wire just exactly as much as would a d-c of one ampere, then we say that the a-c has an "effective value" of one ampere. Of course part of the time of each cycle the stream is larger than an ampere but for part it is less. If the average heating effect is the same the a-c is said to be one ampere.

In the same way, if a steady e. m. f. (a d-c e. m. f.) of one volt will heat a wire to which it is applied a certain amount and if an alternating e. m. f. will have the same heating effect in the same time, then the a-c e. m. f. is said to be one volt.

Another electromagnetic instrument which we have discussed but of which more should be said is the iron-cored transformer. We consider first what happens in one of the coils of the transformer.

The inductance of a coil is very much higher if it has an iron core. The reason is that then the coil acts as if it had an enormously larger number of turns. All the atomic loops of the core add their effects to the loops of the coil. When the current starts it must line up a lot of these atomic loops. When the current stops and these loops turn back 209into some of their old self-satisfied groupings, they affect the electrons in the coil. Where first they opposed the motion of these electrons, now they insist on its being continued for a moment longer. I'll prove that by describing two simple experiments; and then we'll have the basis for understanding the effect of an iron core in a transformer.

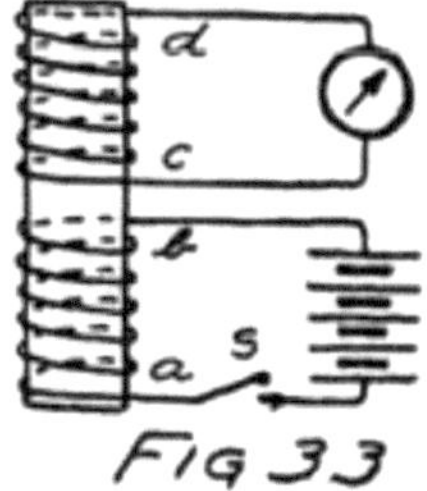

Look again at Fig. 33 of Letter 9 which I am reproducing for convenience. We considered only what would happen in coil *cd* if a current was started in coil *ab*. Suppose instead of placing the coils as shown in that figure they are placed as in Fig. 108. Because they are at right angles there will be no effect in *cd* when the current is started in *ab*. Let the current flow steadily through *ab* and then suddenly turn the coils so that they are again parallel as shown by the dotted positions. We get the same temporary current in *cd* as we would if we should place the coils parallel and then start the current in *ab*.

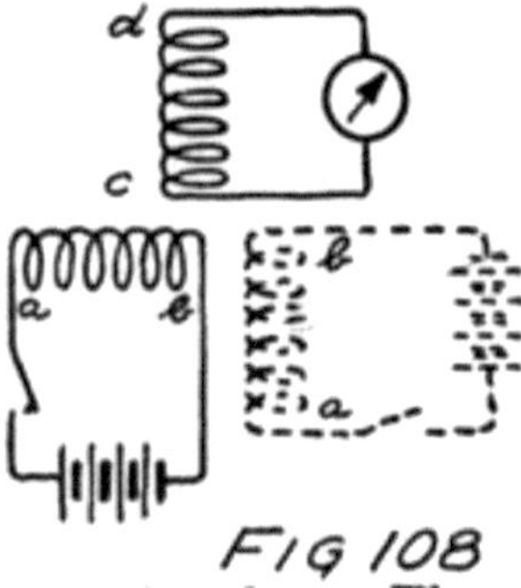

FIG 108

The other experiment is this: Starting with the coils lined up as in the dotted position of Fig. 108 and the current steadily flowing in *ab*, we suddenly turn them into positions at right angles to each other. There is the same momentary current in *cd* as if we had 210 left them lined up and had opened the switch in the circuit of *ab*.

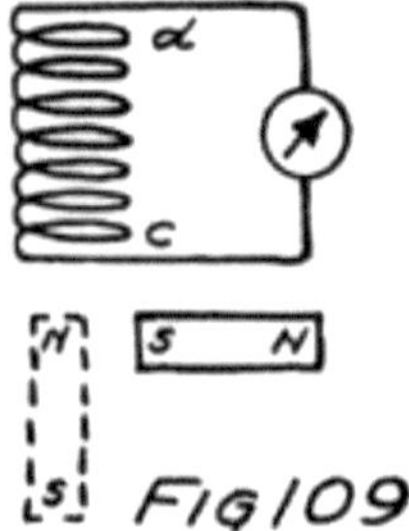

FIG 109

Now we know that the atomic loops of iron behave in the same general way as do loops of wire which are carrying currents. Let us replace the coil *ab* by a magnet as shown in Fig. 109. First we start with the magnet at right angles to the coil *cd*. Suddenly we turn it into the dotted position of that figure. There is the same momentary current in *cd* as if we were still using the coil *ab* instead of a magnet. If now we turn the magnet back to a position at right angles to *cd*,

we observe the opposite direction of current in *cd*. These effects are
more noticeable the more rapidly we turn the magnet. The same is
true of turning the coil.

The experiment of turning the magnet illustrates just what happens in the case of a transformer with, an iron core except that instead of turning the entire magnet the little atomic loops do the
turning inside the core. In the secondary of an iron-cored transformer the induced current is the sum of two currents both in the
same direction at each instant. One current is caused by the starting
or stopping of the current in the primary. The other current is due
to the turning of the atomic loops of the iron atoms so that more of
them line up with the turns of the primary. These atomic loops, of
course, are turned by the current in the primary. There are so
211many of them, however, that the current due to their turning is
usually the more important part of the total current.

In all transformers the effect is greater the more rapidly the current changes direction and the atomic loops turn around. For the
same size of electron stream in the primary, therefore, there is induced in the secondary a greater e. m. f. the greater is the frequency
with which the primary current alternates.

Where high frequencies are dealt with it isn't necessary to have
iron cores because the effect is large enough without the help of the
atomic loops. And even if we wanted their help it wouldn't be easy
to obtain, for they dislike to turn so fast and it takes a lot of power
to make them do so. We know that fact because we know that an
iron core increases the inductance and so chokes the current. For
low frequencies, however, that is those frequencies in the audio
range, it is usually necessary to have iron cores so as to get enough
effect without too many turns of wire.

The fact that iron decreases the inductance and so seriously impedes alternating currents leads us to use iron-core coils where we
want high inductance. Such coils are usually called "choke coils" or
"retard coils." Of their use we shall see more in a later letter where
we study radio-telephone transmitters.

 LETTER 21
YOUR RECEIVING SET AND HOW TO EXPERIMENT

My Dear Student:

In this letter I want to tell you how to experiment with radio apparatus. The first rule is this: Start with a simple circuit, never add anything to it until you know just why you are doing so, and do not box it up in a cabinet until you know how it is working and why.

Your antenna at the start had better be a single wire about 25 feet high and about 75 feet long. This antenna will have capacity of about 0.0001 m. f. If you want an antenna of two wires spaced about three feet apart I would make it about 75 feet long. Bring down a lead from each wire, twisting them into a pigtail to act like one wire except near the horizontal part of the antenna.

FIG 110

Your ground connection can go to a water pipe. To protect the house and your apparatus from lightning insert a fuse and a little carbon block lightning arrester such as are used by the telephone company in their installations of house phones. You can also use a so-called "vacuum lightning arrester." In either 213case the connections will be as shown in Fig. 111. If you use a loop antenna, of course, no arrester is needed.

At first I would plan to receive signals between 150 meters and 360 meters. This will include the amateurs who work between 160 and 200 m., the special amateurs who send C-W telegraph at 275 m., and the broadcasting stations which operate at 360 m. This range will give you plenty to listen to while you are experimenting. In addition you will get some ship signals at 300 m.

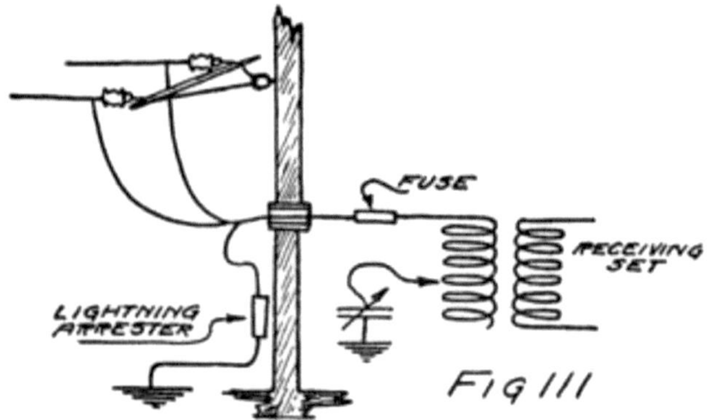

To tune the antenna to any of the wave lengths in this range you can use a coil of 75 turns wound on a cardboard tube of three and a half inches in diameter. You can wind this coil of bare wire if you are careful, winding a thread along with the wire so as to keep the successive turns separated. In that case you will need to construct a sliding contact for it. That is the simplest form of tuner.

On the other hand you can wind with single silk covered wire and bring out taps at the 0, 2, 4, 6, 8, 21410, 14, 20, 28, 36, 44, 56, 66, and 75th turns. To make a tap drill a small hole through the tube, bend the wire into a loop about a foot long and pull this loop through the hole as shown in Fig. 110. Then give the wire a twist, as shown, so that it can't pull out, and proceed with your winding.

Use 26 s. s. c. wire. You will need about 80 feet and might buy 200 to have enough for the secondary coil. Make contacts to the taps by two rotary switches as shown in Fig. 112. You can buy switch arms and contacts studs or a complete switch mounted on a small panel of some insulating compound. Let switch S_1 make the contacts for taps between 14 and 75 turns, and let switch S_2 make the other contacts.

For the secondary coil use the same size of wire and of core. Wind 60 turns, bringing out a tap at the middle. To tune the secondary circuit you will need a variable condenser. You can buy one of the small ones with a maximum capacity of about 0.0003 mf., one of the larger ones with a maximum capacity of 0.0005 mf., or even the larger size which has a maximum capacity of 0.001 mf. I should prefer the one of 0.0005 mf.

You will need a crystal detector–I should try galena first–and a so-called "cat's whisker" with which to make contact with the galena. For these parts and for the switch mentioned above you can shop around to advantage. For telephone receivers I would buy a really good pair with a resistance of about 2500 ohms. Buy also a small mica condenser 215of 0.002 mf. for a blocking condenser. Your entire outfit will then look as in Fig. 112. The switch S is a small knife switch.

To operate, leave the switch S open, place the primary and secondary coils near together as in the figure and listen. The tuning is varied, while you listen, by moving the slider of the slide-wire tuner or by moving the switches if you have connected your coil for that method. Make large changes in the tuning by varying the switch S_1 and then turn slowly through all positions of S_2 , listening at each position.

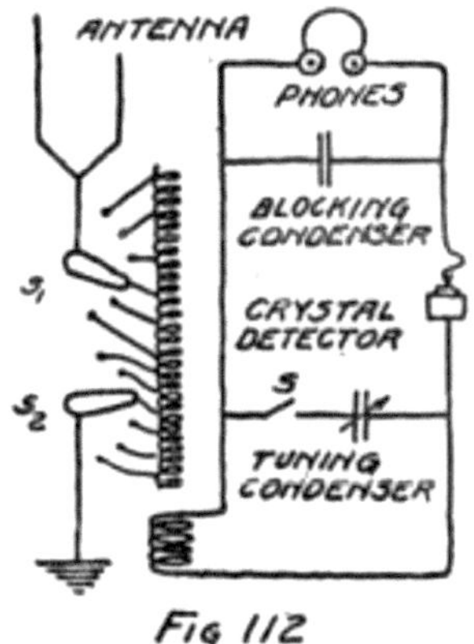

Fig 112

When a signal is heard adjust to the position of S_1 and S_2 which gives the loudest signal and then closing S start to tune the secondary circuit. To do this, vary the capacity of the condenser in the secondary circuit. Don't change the primary tuning until you have tuned the secondary and can get the signal with good volume, that is loud. You will want to vary the position of the primary and secondary coils, that is, vary their coupling, for you will get sharper tuning as they are drawn farther apart. Sharper tuning means less interference from other stations which are sending on wave lengths near that which you wish to receive. Reduce the coupling, therefore, and then readjust the tuning. It will usually be necessary to make a slight change in both circuits, in one 216 case with switch S_1 and in the other with the variable condenser.

As soon as you can identify any station which you hear sending make a note of the position of the switches S_1 and S_2 , and of the setting of the condenser in the secondary circuit. In that way you will acquire information as to the proper adjustments to receive certain wave-lengths. This is calibrating your set by the known wave-lengths of distant stations.

After learning to receive with this simple set I should recommend buying a good audion tube. Ask the seller to supply you with a blue print of the characteristic [11] of the tube taken under the conditions of filament current and plate voltage which he recommends for its use. Buy a storage battery and a small slide-wire rheostat, that is variable resistance, to use in the filament circuit. Buy also a bank of dry batteries of the proper voltage for the plate circuit of the tube. At the same time you should buy the proper design of transformer to go between the plate circuit of your tube and the pair of receivers which you have. It will usually be advisable to ask the dealer to show you a characteristic curve for the transformer, which will indicate how well the transformer operates at the different frequencies in the audio range. It should operate very nearly the same for all frequencies between 200 and 2500 cycles.

The next step is to learn to use the tube as a 217detector. Connect it into your secondary circuit instead of the crystal detector. Use the proper value of C-battery as determined from your study of the characteristic of the tube. One or two small dry cells, which have binding-post terminals are convenient C-batteries. If you think you will need a voltage much different from that obtained with a whole number of batteries you can arrange to supply the grid as we did in Fig. 86 of Letter 18. In that case you can use a few feet of 30 German-silver wire and make connections to it with a suspender clip. Learn to receive with the tube and be particularly careful not to let the filament have too much current and burn out.

Now buy some more apparatus. You will need a grid condenser of about 0.0002 mf. The grid leaks to go with it you can make for yourself. I would use a piece of brown wrapping paper and two little metal eyelets. The eyelets can be punched into the paper. Between them coat the paper with carbon ink, or with lead pencil marks. A line about an inch long ought to serve nicely. You will probably wish to make several grid leaks to try. When you get satisfactory operation in receiving by the grid-condenser method the leak will probably be somewhere between a megohm and two megohms.

For this method you will not want a C-battery, but you will wish to operate the detector with about as high a voltage as the manufacturers will recommend for the plate circuit. In this way the incoming signal, which decreases the plate current, can 218produce the largest decrease. It is also possible to start with the grid slightly positive instead of being as negative as it is when connected to the negative terminal of the A-battery. There will then be possible a greater change in grid voltage. To do so connect the grid as in Fig. 115 to the positive terminal of the A-battery.

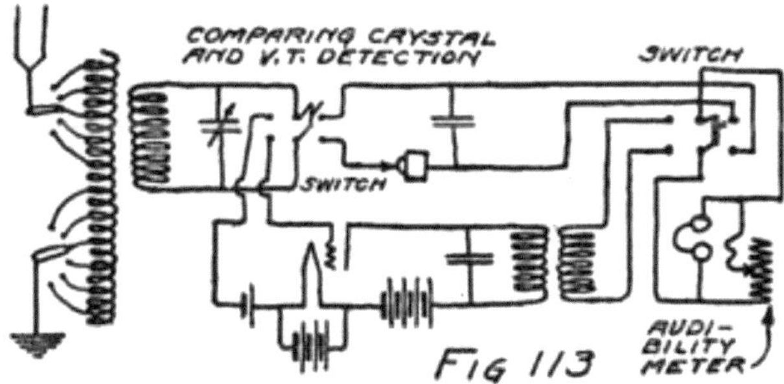

About this time I would shop around for two or three small double-pole double-throw switches. Those of the 5-ampere size will do. With these you can arrange to make comparisons between different methods of receiving. Suppose, for example, you connect the switches as shown in Fig. 113 so that by throwing them to the left you are using the audion and to the right the crystal as a detector. You can listen for a minute in one position and then switch and listen for a minute in the other position, and so on back and forth.

That way you can tell whether or not you really are getting better results.

If you want a rough measure of how much better the audion is than the crystal you might see, while you are listening to the audion, how much you can rob the telephone receiver of its current and still hear as well as you do when you switch back to the crystal. The easiest way to do this is to put a variable resistance across the receiver as shown in Fig. 113. Adjust this resistance until the intensity of the signal when detected by the audion is the same as for the crystal. You adjust this variable resistance until it by-passes so much of the current, which formerly went through the receiver, that the "audibility" of the signal is reduced until it is the same as for the crystal detector. Carefully made resistances for such a purpose are sold under the name of "audibility meters." You can assemble a resistance which will do fairly well if you will buy a small rheostat which will give a resistance varying by steps of ten ohms from zero to one hundred ohms. At the same time you can buy four resistance spools of one hundred ohms each and perhaps one of 500 ohms. The spools need not be very expensive for you do not need carefully adjusted resistances. Assemble them so as to make a rheostat with a range of 0-1000 ohms by steps of 10 ohms. The cheapest way to mount is with Fahnestock clips as illustrated in Fig. 114. After a while, however, you will probably wish to mount them in a box with a rotary switch on top.

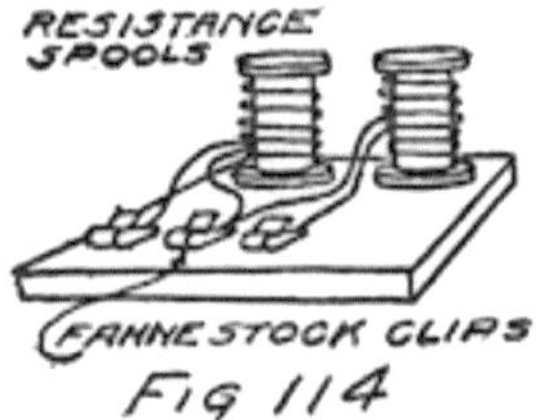

To study the effect of the grid condenser you can arrange switches so as to insert this condenser and its leak and at the same time to cut out the C-battery. Fig. 115 shows how. You can measure the gain in audibility at the same time.

Pl. X.–Audio-frequency Transformer and Banked-wound Coil.
(Courtesy of Pacent Electric Co.)

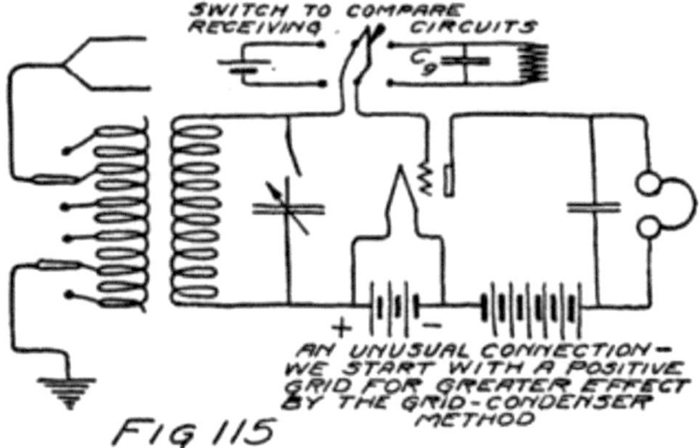

219After learning to use the audion as a detector, both by virtue of its curved characteristic and by the grid-condenser method, I would suggest studying the same tube as an amplifier. First I would learn to use it as an audio-frequency amplifier. Set up the crystal detector circuit. Use your audio-frequency transformer the other way around so as to step up to the grid. Put the telephone in the plate circuit. Choose your C-battery for amplification and *not detection* and try to receive.

You will get better results if you can afford another iron-core transformer. If you can, buy one which will work between the plate circuit of one vacuum tube and the grid circuit of another similar tube. Then you will have the right equipment when you come to make a two-stage audio-frequency amplifier. If you buy such a transformer use the other transformer between plate and telephones as you did before and insert the new one as shown in Fig. 116. 220This circuit also shows how you can connect the switches so as to see how much the audion is amplifying.

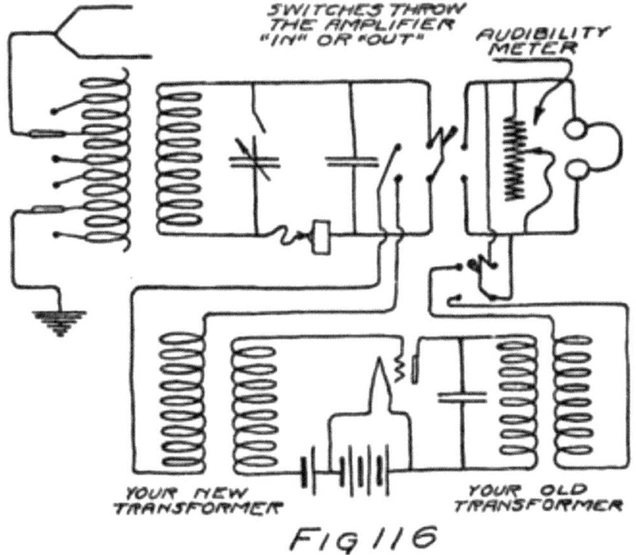

FIG 116

The next step is to use the audion as an amplifier of the radio-signal before its detection. Use the proper C-battery for an amplifier, as determined from the blue print of the tube characteristic. Connect the tube as shown in Fig. 117. You will see that in this circuit we are using a choke coil to keep the radio-frequency current out of the battery part of the plate circuit and a small condenser, another one of 0.002 mf., to keep the battery current from the crystal detector. You can see from the same figure how you can arrange the switches so as to find whether or not you are getting any gain from the amplifier.

221Now you are ready to receive those C-W senders at 275 meters. You will need to wind another coil like the secondary coil you already have. Here is where you buy another condenser. You will need it later. If before you bought the 0.0005 size, this time buy the 0.001 size or vice versa. Wind also a small coil for a tickler. About 20 turns of 26 wire on a core of 3-1/2 in. diameter will do. Connect the tickler in the plate circuit of the audion. Connect to the grid your new coil and condenser and set the audion circuit so that it will induce a current in the secondary circuit which supplies the crystal. Fig. 118 shows the hook-up.

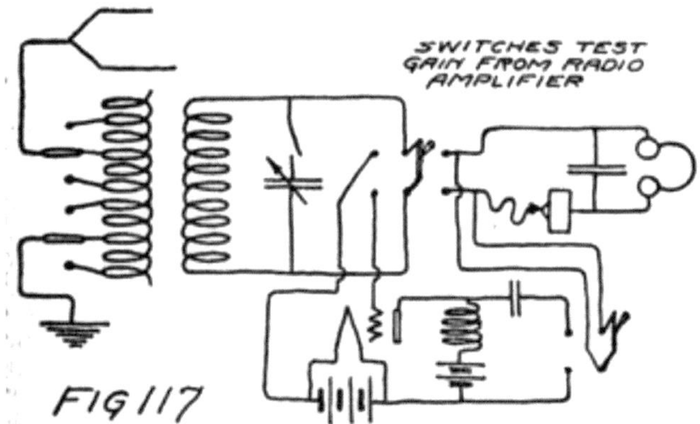

You will see that you are now supplying the crystal with current from two sources, namely the distant source of the incoming signals and the local oscillator which you have formed. The crystal will detect the "beat note" between these two currents.

To receive the 275 meters signals you will need to make several adjustments at the same time. In the first place I would set the tuning of the antenna 222circuit and of the crystal circuit about where you think right because of your knowledge of the settings for other wave lengths. Then I would get the local oscillator going. You can tell whether or not it is going if you suddenly increase or decrease the coupling between the tickler coil and the input circuit of the audion. If this motion is accompanied by a click in the receivers the tube is oscillating.

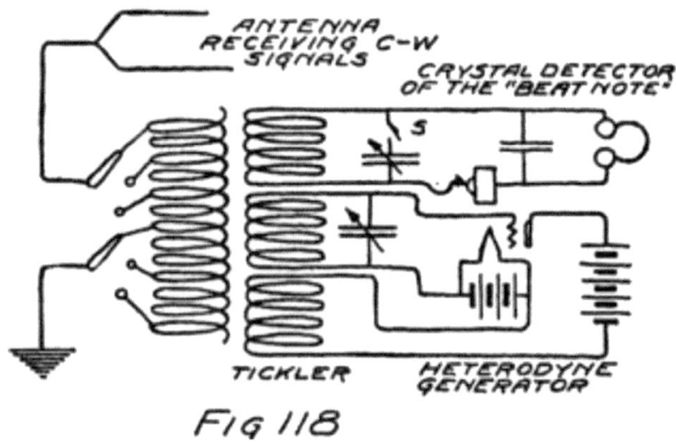

Now you must change the frequency at which it is oscillating by slowly changing the capacity in the tuned input circuit of the tube. Unless the antenna circuit is properly tuned to the 275 meter signal you will get no results. If it is, you will hear an intermittent musical note for some tune of your local oscillator. This note will have the duration of dots and dashes.

You will have to keep changing the tuning of your detector circuit and of the antenna. For each new setting very slowly swing the condenser plates in the oscillator circuit and see if you get a signal. It 223will probably be easier to use the "stand-by position," which I have described, with switch S open in the secondary circuit of Fig. 118. In that case you have only to tune your antenna to 275 meters and then you will pick up a note when your local oscillator is in tune. After you have done so you can tune the secondary circuit which supplies the crystal.

If you adopt this method you will want a close coupling between the antenna and the crystal circuit. You will always want a very weak coupling between the oscillator circuit and the detector circuit. You will also probably want a weaker coupling between tickler and tube input than you are at first inclined to believe will be enough. Patience and some skill in manipulation is always required for this sort of experiment.

When you have completed this experiment in heterodyne receiving, using a local oscillator, you are ready to try the regenerative circuit. This has been illustrated in Fig. 92 of Letter 18 and needs no further description. You will have the advantage when you come to this of knowing very closely the proper settings of the antenna circuit and the secondary tuned circuit. You will need then only to adjust the coupling of the tickler and make finer adjustments in your tuning.

After you have completed this series of experiments you will be something of an adept at radio and are in a position to plan your final set. For this set you will need to purchase certain parts 224complete from reputable dealers because many of the circuits which I have described are patented and should not be used except as rights to use are obtained by the purchase of licensed apparatus which embodies the patented circuits. Knowing how radio receivers operate and why, you are now in a good condition to discuss with dealers the relative merits and costs of receiving sets.

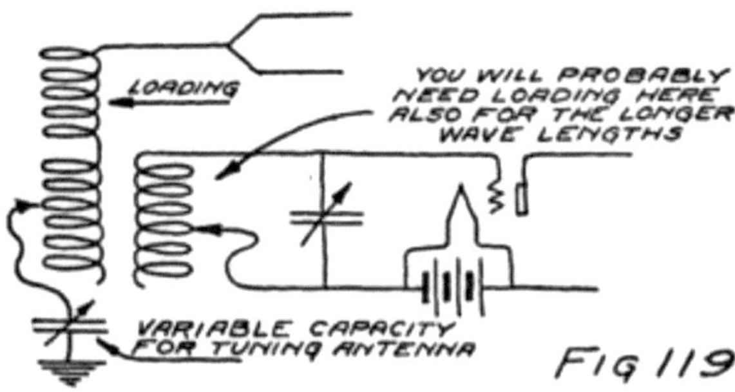

Before you actually buy a completed set you may want to increase the range of frequency over which you are carrying out your experiments. To receive at longer wave-lengths you will need to increase the inductance of your antenna so that it will be tuned to a lower frequency. This is usually called "loading" and can be done by inserting a coil in the antenna. To obtain smaller wave-lengths decrease the effective capacity of the antenna circuit by putting another condenser in series with the antenna. Usually, therefore, one connects into his antenna circuit both a condenser and a loading coil. By using a variable condenser the effective capacity of the antenna system may be easily changed. The result is that this 225series condenser method becomes the easiest method of tuning and the slide wire tuner is not needed. Fig. 119 shows the circuit.

For quite a range of wave-lengths we may use the same loading coil and tune the antenna circuit entirely by this series condenser. For some other range of wave-lengths we shall then need a different loading coil. In a well-designed set the wave-length ranges overlap. The calculation of the size of loading coil is quite easy but requires more arithmetic than I care to impose on you at present. I shall therefore merely give you illustrations based on the assumption that your antenna has a capacity of 0.0001 or of 0.0002 mf. and that the condensers which you have bought are 0.0005 and 0.001 for their maxima.

In Table I there is given, for each of several values of the inductance of the primary coil, the shortest and the longest wave-lengths which you can expect to receive. The table is in two parts, the first for an antenna of capacity 0.0001 mf. and the second for one of 0.0002 mf. Yours will be somewhere between these two limits. The shortest wave-length depends upon the antenna and not upon the condenser which you use in series with it for tuning. It also depends upon how much inductance there is in the coil which you have in the antenna circuit. The table gives values of inductance in the first column, and of minimum wave-length in the second. The third column shows what is the greatest wave-length you may expect if you use a tuning condenser of 0.0005 mf.; and the fourth column the slightly 226large wave-length which is possible with the larger condenser.

TABLE I

Part 1. (For antenna of 0.0001 mf.)

Inductance in mil-henries	Shortest wave-length in meters.	Longest wave-length in meters with 0.0005 mf.	Longest wave-length in meters with 0.001 mf.
0.10	103	169	179
0.20	146	238	253
0.40	207	337	358
0.85	300	490	515
1.80	400	700	760
2.00	420	750	800
4.00	600	1080	1130
5.00	660	1200	1260
10.00	900	1700	1790
30.00	1600	2900	3100

Part 2. (For antenna of 0.0002 mf.)

0.10	169	225	240
0.16	210	285	305
0.20	240	320	340
0.25	270	355	380

0.40	340	450	480
0.60	420	550	590
0.80	480	630	680
1.20	585	775	840
1.80	720	950	1020
3.00	930	1220	1320
5.00	1200	1600	1700
8.00	1500	2000	2150
12.00	1850	2400	2650
16.00	2150	2800	3050

From Table I you can find how much inductance you will need in the primary circuit. A certain amount you will need to couple the antenna and the secondary circuit. The coil which you wound at the beginning of your experiments will do well for that. Anything more in the way of inductance, which the antenna circuit requires to give a desired wave-length, you may consider as loading. In Table II are some data as to winding coils on straight 227cores to obtain various values of inductance. Your 26 s. s. c. wire will wind about 54 turns to the inch. I have assumed that you will have this number of turns per inch on your coils and calculated the inductance which you should get for various numbers of total turns. The first part of the table is for a core of 3.5 inches in diameter and the second part for one of 5 inches. The first column gives the inductance in mil-henries. The second gives number of turns. The third and fourth are merely for convenience and give the approximate length in inches of the coil and the approximate total length of wire which is required to wind it. I have allowed for bringing out taps. In other words 550 feet of the wire will wind a coil of 10.2 inches with an inductance of 8.00 mil-henries, and permit you to bring out taps at all the lower values of inductance which are given in the table.

TABLE II

Part 1. (For a core of 3.5 in. diam.)

Inductance in mil-henries	Number of turns	Length in inches	Feet of wire required.

0.10	25	0.46	25
0.16	34	0.63	36
0.20	39	0.72	42
0.25	44	0.81	49
0.40	58	1.07	63
0.60	75	1.38	80
0.80	92	1.70	100
0.85	96	1.78	104
1.00	108	2.00	118
1.20	123	2.28	133
1.80	164	3.03	176
2.00	180	3.33	190
3.00	242	4.48	250
4.00	304	5.62	310
5.00	366	6.77	370
8.00	550	10.20	550

Part 2. (For a core of 5.0 in. diam.)

2.00	120	2.22	160
3.00	158	2.93	215
4.00	194	3.58	265
5.00	228	4.22	310
8.00	324	6.00	450
10.00	384	7.10	530
12.00	450	8.30	625

228The coil which you wound at the beginning of your experiment had only 75 turns and was tapped so that you could, by manipulating the two switches of Fig. 112, get small variations in inductance. In Table III is given the values of the inductance which is controlled by the switches of that figure, the corresponding number of turns, and the wave-length to which the antenna should then be tuned. I am giving this for two values of antenna capacity, as I have done before. By the aid of these three tables you should have small difficulty in taking care of matters of tuning for all wave-lengths below about 3000 meters. If you want to get longer waves than that you had better buy a few banked-wound coils. These are coils in which the turns are wound over each other but in such a way as to avoid in large part the "capacity effects" which usually accompany such winding. You can try winding them for yourself but I doubt if the experience has much value until you have gone farther in the study of the mathematical theory of radio than this series of letters will carry you.

TABLE III

Number of turns	Inductance in mil-henries	Circuit of Fig. 112 Wavelength with antenna of	
		0.0001 mf.	0.0002 mf.
14	0.04	120	170
20	0.07	160	220
28	0.12	210	290
36	0.18	250	360
44	0.25	300	420
56	0.38	370	520
75	0.60	460	650

In the secondary circuit there is only one capacity, that of the variable condenser. If it has a range of values from about 0.00005 mf. to 0.0005 mf. your coil of 60 turns and 0.42 mf. permits a range of wave-lengths from 270 to 860 m. Using half the coil the range is 150 to 480 m. With the larger condenser the ranges are respectively 270 to 1220 and 270 to 670. For longer wave-lengths load with inductance. Four times the inductance will tune to double these wave-lengths.

[11]

If you can afford to buy, or if you can borrow, ammeters and voltmeters of the proper range you should take the characteristic yourself.

230 LETTER 22
HIGH-POWERED RADIO-TELEPHONE TRANS-MITTERS

My Dear Experimenter:

This letter is to summarize the operations which must be performed in radio-telephone transmission and reception; and also to describe the circuit of an important commercial system.

To transmit speech by radio three operations are necessary. First, there must be generated a high-frequency alternating current; second, this current must be modulated, that is, varied in intensity in accordance with the human voice; and third, the modulated current must be supplied to an antenna. For efficient operation, of course, the antenna must be tuned to the frequency which is to be transmitted. There is also a fourth operation which is usually performed and that is amplification. Wherever the electrical effect is smaller than desired, or required for satisfactory transmission, vacuum tubes are used as amplifiers. Of this I shall give you an illustration later.

Three operations are also essential in receiving. First, an antenna must be so arranged and tuned as to receive energy from the distant transmitting station. There is then in the receiving antenna a current similar in wave form to that in the transmitting 231antenna. Second, the speech significance of this current must be detected, that is, the modulated current must be demodulated. A current is then obtained which has the same wave form as the human voice which was the cause of the modulation at the distant station. The third operation is performed by a telephone receiver which makes the molecules of air in its neighborhood move back and forth in accordance with the detected current. As you already know a fourth operation may be carried on by amplifiers which give on their output sides currents of greater strength but of the same forms as they receive at their input terminals.

In transmitting and in receiving equipment two or more of these operations may be performed by the same vacuum tube as you will remember from our discussion of the regenerative circuit for receiving. For example, also, in any receiving set the vacuum tube which detects is usually amplifying. In the regenerative circuit for receiving continuous waves by the heterodyne method the vacuum tube functions as a generator of high-frequency current and as a detector of the variations in current which occur because the locally-generated current does not keep in step with that generated at the transmitting station.

Another example of a vacuum tube performing simultaneously two different functions is illustrated in Fig. 120 which shows a simple radio-telephone transmitter. The single tube performs in itself both the generation of the radio-frequency current and its 232modulation in accordance with the output of the carbon-button transmitter. This audion is in a feed-back circuit, the oscillation frequency of which depends upon the condenser C and the inductance L. The voice drives the diaphragm of the transmitter and thus varies the resistance of the carbon button. This varies the current from the battery, B_A, through the primary, T_1, of the transformer T. The result is a varying voltage applied to the grid by the secondary T_2. The oscillating current in the plate circuit of the audion varies accordingly because it is dependent upon the grid voltage. The condenser C_R offers a low impedance to the radio-frequency current to which the winding T_2 of audio-frequency transformer offers too much.

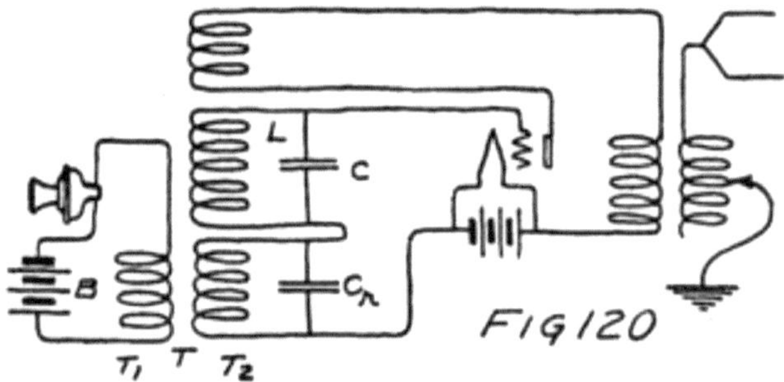

In this case the tube is both generator and "modulator." In some cases these operations are separately performed by different tubes. This was true of the transmitting set used in 1915 when the engineers of the Bell Telephone System talked by radio from Arlington, near Washington, D. C., to Paris and Honolulu. I shall not draw out completely the circuit of their apparatus but I shall describe it by

233using little squares to represent the parts responsible for each of the several operations.

First there was a vacuum tube oscillator which generated a small current of the desired frequency. Then there was a telephone transmitter which made variations in a direct-current flowing through the primary of a transformer. The e. m. f. from the secondary of this transformer and the e. m. f. from the radio-frequency oscillator were both impressed upon the grid of an audion which acted as a modulator. The output of this audion was a radio-frequency current modulated by the voice. The output was amplified by a two-stage audion amplifier and supplied through a coupling coil to the large antenna of the U. S. Navy Station at Arlington. Fig. 121 shows the system.

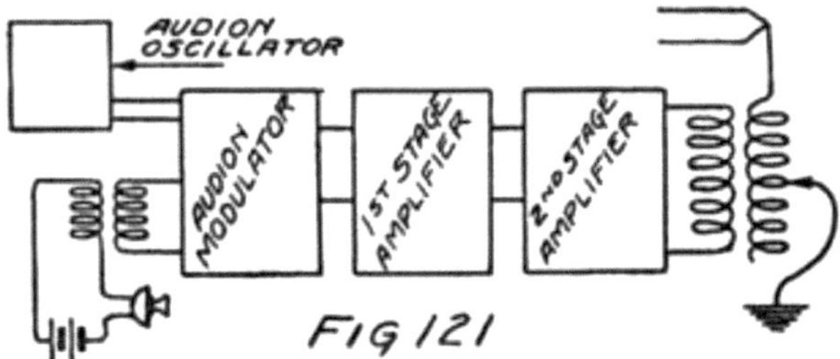

The audion amplifiers each consisted of a number of tubes operating in parallel. When tubes are operated in parallel they are connected as shown in Fig. 122 so that the same e. m. f. is impressed on all the grids and the same plate-battery voltage on all the plates. As the grids vary in voltage there is a corresponding variation of current in the plate circuit of each tube. The total change of the current 234in the plate-battery circuit is, then, the sum of the changes in all the plate-filament circuits of the tubes. This scheme of connections gives a result equivalent to that of a single tube with a correspondingly larger plate and filament.

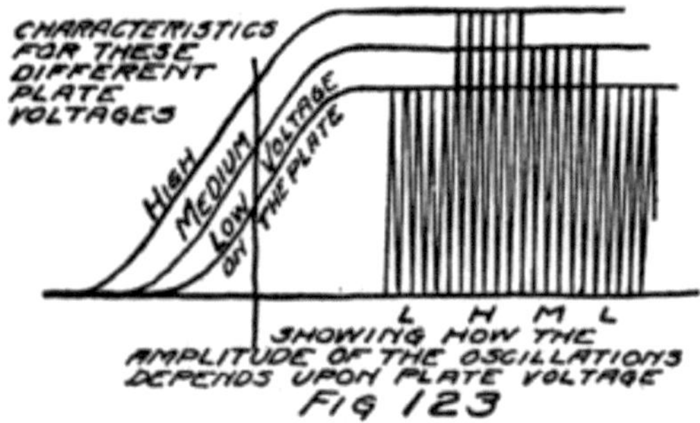

Parallel connection is necessary because a single tube would be overheated in delivering to the antenna the desired amount of power. You remember that when the audion is operated as an amplifier the resistance to which it supplies current is made equal to its own internal resistance of R_P . That means that there is in the plate circuit just as much resistance inside the tube as outside. Hence there is the same amount of work done each second in forcing the current through the tube as through the antenna circuit, if that is what the tube supplies. "Work per second" is power; the plate battery is spending energy in the tube at the same rate as it is supplying it to the antenna where it is useful for radiation.

Pl. XI.–Broadcasting Equipment, Developed by the American Telephone and Telegraph Company and the Western Electric Company.

235All the energy expended in the tube appears as heat. It is due to the blows which the electrons strike against the plate when they are drawn across from the filament. These impacts set into more rapid motion the molecules of the plate; and the temperature of the tube rises. There is a limit to the amount the temperature can rise without destroying the tube. For that reason the heat produced inside it must not exceed a certain limit depending upon the design of the tube and the method of cooling it as it is operated. In the Arlington experiments, which I mentioned a moment ago, the tubes were cooled by blowing air on them from fans.

We can find the power expended in the plate circuit of a tube by multiplying the number of volts in its battery by the number of amperes which flows. Suppose the battery is 250 volts and the current 0.02 amperes, then the power is 5 watts. The "watt" is the unit for measuring power. Tubes are rated by the number of watts which can be safely expended in them. You might ask, when you buy an audion, what is a safe rating for it. The question will not be an important one, however, unless you are to set up a transmitting set since a detector is usually operated with such small plate-voltage as not to have expended in it an amount of power dangerous to its life.

In recent transmitting sets the tubes are used in parallel for the reasons I have just told, but a different 236method of modulation is used. The generation of the radio-frequency current is by large-powered tubes which are operated with high voltages in their plate circuits. The output of these oscillators is supplied to the antenna. The intensity of the oscillations of the current in these tubes is controlled by changing the voltage applied in their plate circuits. You can see from Fig. 123 that if the plate voltage is changed the strength of the alternating current is changed accordingly. It is the method used in changing the voltage which is particularly interesting.

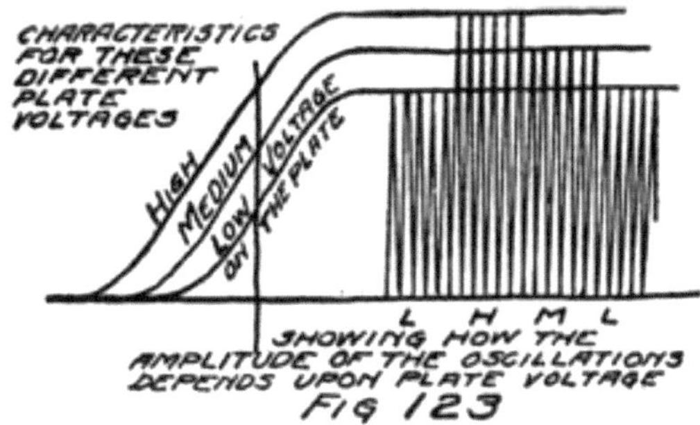

The high voltages which are used in the plate circuits of these high-powered audions are obtained from generators instead of batteries. You remember from Letter 20 that an e. m. f. is induced in a coil when the coil and a magnet are suddenly changed in their positions, one being turned with reference to the other. A generator is a machine for turning a coil so that a magnet is always inducing an e. m. f. in it. It is formed by an armature carrying coils and by strong electromagnets. The machine can be 237driven by a steam or gas engine, by a water wheel, or by an electric motor. Generators are designed either to give steady streams of electrons, that is for d-c currents, or to act as alternators.

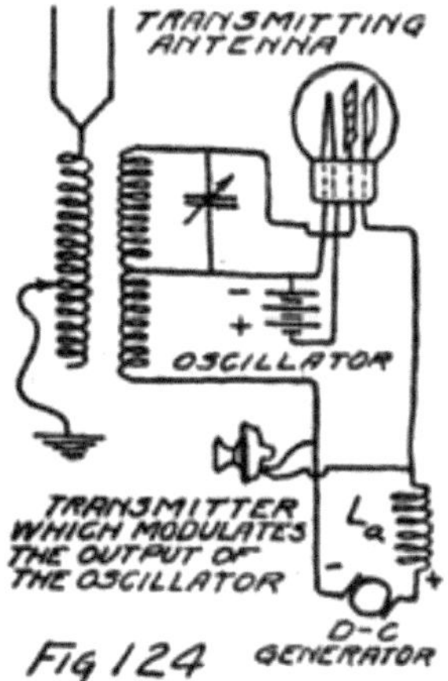

Suppose we have, as shown in Fig. 124, a d-c generator supplying current to a vacuum tube oscillator. The current from the generator passes through an iron-cored choke coil, marked L_A in the figure.

Between this coil and the plate circuit we connect across the line a telephone transmitter. To make a system which will work efficiently we shall have to suppose that this transmitter has a high resistance, say about the same as the internal resistance, R_P , of the tube and also that it can carry as large a current.

Of the current which comes from the generator about one-half goes to the tube and the rest to the transmitter. If the resistance of the transmitter is increased it can't take as much current. The coil, L_A , however, because of its inductance, tends to keep the same amount of current flowing through itself. For just an instant then the current in L_A keeps steady even though the transmitter doesn't take its share. The result is more current for the oscillating tube. On the other hand if the transmitter takes more current, because its resistance is decreased, 238the choke coil, L_A , will momentarily tend to keep the current steady so that what the transmitter takes must be at the expense of the oscillating tube.

That's one way of looking at what happens. We know, however, from Fig. 123 that to get an increase in the amplitude of the current in the oscillating tube we must apply an increased voltage to its plate circuit. That is what really happens when the transmitter increases in resistance and so doesn't take its full share of the current. The reason is this: When the transmitter resistance is increased the current in the transmitter decreases. Just for a moment it looks as though the current in L_A is going to decrease. That's the way it looks to the electrons; and you know what electrons do in an inductive circuit when they think they shall have to stop. They induce each other to keep on for a moment. For a moment they act just as if there was some extra e. m. f. which was acting to keep them going. We say, therefore, that there is an extra e. m. f., and we call this an e. m. f. of self-induction. All this time there has been active on the plate circuit of the tube the e. m. f. of the generator. To this there is added at the instant when the transmitter resistance increases, the e. m. f. of self-induction in the coil, L_A ; and so the total e. m. f. applied to the tube is momentarily increased. This increased e. m. f., of course, results in an increased amplitude for the alternating current which the oscillator is supplying to the transmitting antenna.

When the transmitter resistance is decreased, and 239a larger current should flow through the choke coil, the electrons are asked to speed up in going through the coil. At first they object and during that instant they express their objection by an e. m. f. of self-induction which opposes the generator voltage. For an instant, then, the voltage of the oscillating tube is lowered and its alternating-current output is smaller.

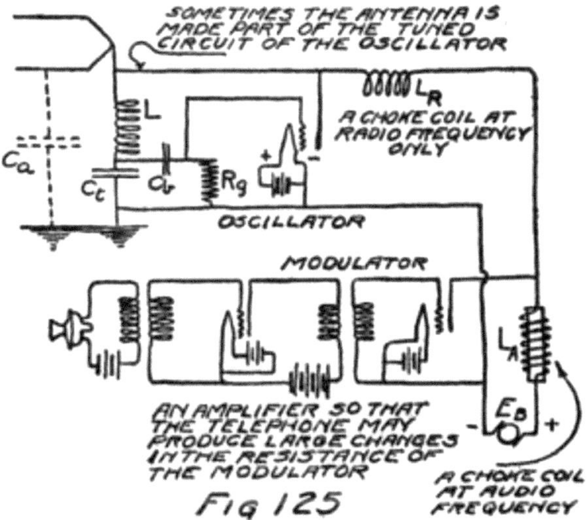

FIG 125

For the purpose of bringing about such threatened changes in current, and hence such e. m. f.'s of self-induction, the carbon transmitter is not suitable because it has too small a resistance and too small a current carrying ability. The plate circuit of a vacuum tube will serve admirably. You know from the audion characteristic that without changing the plate voltage we can, by applying a voltage to the grid, change the current through the plate circuit. 240Now if it was a wire resistance with which we were dealing and we should be able to obtain a change in current without changing the voltage acting on this wire we would say that we had changed the resistance. We can say, therefore, that the internal resistance of the plate circuit of a vacuum tube can be changed by what we do to the grid.

In Fig. 125 I have substituted the plate circuit of an audion for the transmitter of Fig. 124 and arranged to vary its resistance by changing the potential of the grid. This we do by impressing upon the grid the e. m. f. developed in the secondary of a transformer, to the primary of which is connected a battery and a carbon transmitter. The current through the primary varies in accordance with the sounds spoken into the transmitter. And for all the reasons which we have just finished studying there are similar variations in the output current of the oscillating tube in the transmitting set of Fig. 125.

In this latter figure you will notice a small air-core coil, L_R , between the oscillator and the modulator tube. This coil has a small inductance but it is enough to offer a large impedance to radio-frequency currents. The result is, it does not let the alternating currents of the oscillating tube flow into the modulator. These currents are confined to their own circuit, where they are useful in establishing similar currents in the antenna. On the other hand, the coil L_R doesn't seriously impede low-frequency currents and therefore it does not prevent variations 241 in the current which are at audio-frequency. It does not interfere with the changes in current which accompany the variations in the resistance of the plate circuit of the modulator. That is, it has too little impedance to act like L_A and so it permits the modulator to vary the output of the oscillator.

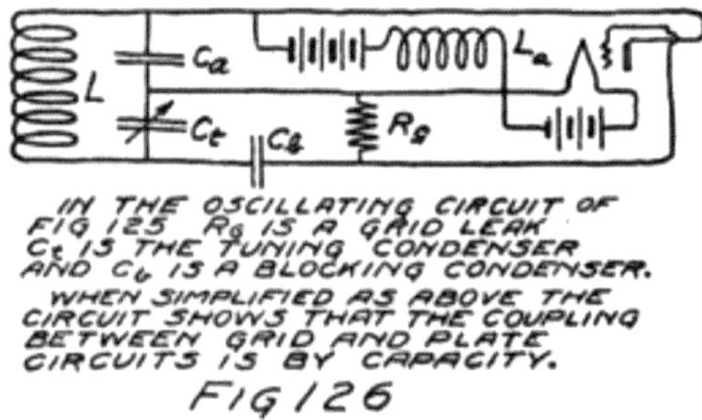

The oscillating circuit of Fig. 125 includes part of the antenna. It differs also from the others I have shown in the manner in which grid and plate circuits are coupled. I'll explain by Fig. 126.

The transmitting set which I have just described involves many of the principles of the most modern sets. If you understand its operation you can probably reason out for yourself any of the other sets of which you will hear from time to time.

242 LETTER 23
AMPLIFICATION AT INTERMEDIATE FRE-QUENCIES

DEAR SON:

In the matter of receiving I have already covered all the important principles. There is one more system, however, which you will need to know. This is spoken of either as the "super-heterodyne" or as the "intermediate-frequency amplification" method of reception.

The system has two important advantages. First, it permits sharper tuning and so reduces interference from other radio signals. Second, it permits more amplification of the incoming signal than is usually practicable.

First as to amplification: We have seen that amplification can be accomplished either by amplifying the radio-frequency current before detection or by amplifying the audio-frequency current which results from detection. There are practical limitations to the amount of amplification which can be obtained in either case. An efficient multi-stage amplifier for radio-frequencies is difficult to build because of what we call "capacity effects."

Consider for example the portion of circuit shown in Fig. 127. The wires a and b act like small plates of condensers. What we really have, is a lot of 243 tiny condensers which I have shown in the figure by the light dotted-lines. If the wires are transmitting high-frequency currents these condensers offer tiny waiting-rooms where the electrons can run in and out without having to go on to the grid of the next tube. There are other difficulties in high-frequency amplifiers. This one of capacity effects between parallel wires is enough for the present. It is perhaps the most interesting because it is always more or less troublesome whenever a pair of wires is used to transmit an alternating current.

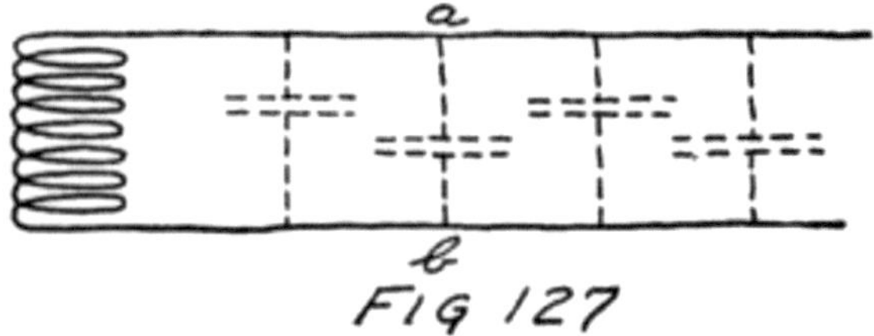

FIG 127

In the case of a multi-stage amplifier of audio-frequency current there is always the possibility of the amplification of any small variations in current which may naturally occur in the action of the batteries. There are always small variations in the currents from batteries, due to impurities in the materials of the plates, air bubbles, and other causes. Ordinarily we don't observe these changes because they are too small to make an audible sound in the telephone receivers. Suppose, however, that they take place in the battery of the first tube of a series of amplifiers. Any tiny change of current is amplified many times and results in a troublesome noise in 244the telephone receiver which is connected to the last tube.

In both types of amplifiers there is, of course, always the chance that the output circuit of one tube may be coupled to and induce some effect in the input circuit of one of the earlier tubes of the series. This will be amplified and result in a greater induction. In other words, in a circuit where there is large amplification, there is always the difficulty of avoiding a feed-back of energy from one tube to another so that the entire group acts like an oscillating circuit, that is "regeneratively." Much of this difficulty can be avoided after experience.

If a multi-stage amplifier is to be built for a current which does not have too high a frequency the "capacity effects" and the other difficulties due to high-frequency need not be seriously troublesome. If the frequency is not too high, but is still well above the audible limit, the noises due to variations in battery currents need not bother for they are of quite low frequency. Currents from 20,000 to 60,000 cycles a second are, therefore, the most satisfactory to amplify.

Suppose, however, one wishes to amplify the signals from a radio-broadcasting station. The wave-length is 360 meters and the frequency is about 834,000 cycles a second. The system of intermediate-frequency amplification solves the difficulty and we shall see how it does so.

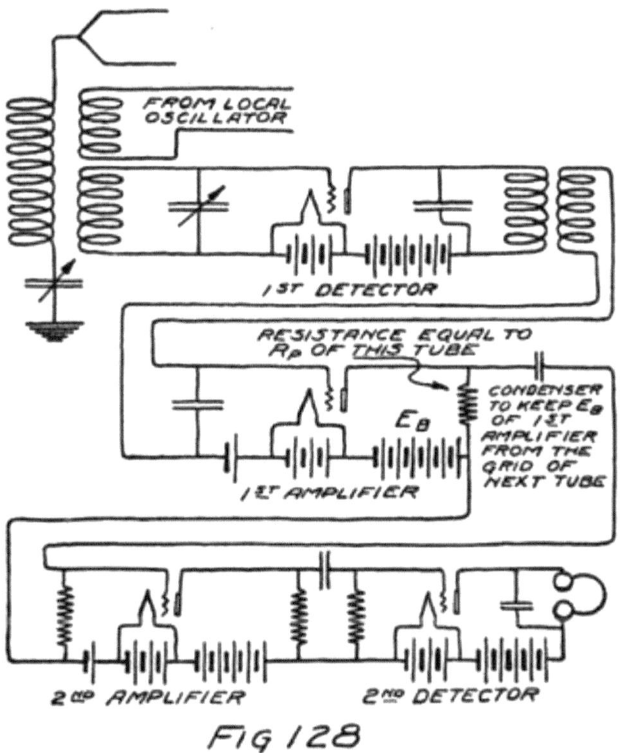

245At the receiving station a local oscillator is used. This generates a frequency which is about 30,000 cycles less than that of the incoming signal. Both currents are impressed on the grid of a detector. The result is, in the output of the detector, a current which has a frequency of 30,000 cycles a second. The intensity of this detected current depends upon the intensity of the incoming signal. The "beat note" current of 30,000 cycles varies, therefore, in accordance with the voice which is modulating at the distant sending station. The speech significance is now hidden in a current of a frequency intermediate between radio and audio. This current may 246be amplified many times and then supplied to the grid of a detector which obtains from it a current of audio-frequency which has a speech significance. In Fig. 128 I have indicated the several operations.

We can now see why this method permits sharper tuning. The whole idea of tuning, of course, is to arrange that the incoming signal shall cause the largest possible current and at the same time to provide that any signals at other wave-lengths shall cause only negligible currents. What we want a receiving set to do is to distinguish between two signals which differ slightly in wave-length and to respond to only one of them.

Suppose we set up a tuned circuit formed by a coil and a condenser and try it out for various frequencies of signals. You know how it will respond from our discussion in connection with the tuning curve of Fig. 51 of Letter 13. We might find from a number of such tests that the best we can expect any tuned circuit to do is to discriminate between signals which differ about ten percent in frequency, that is, to receive well the desired signal and to fail practically entirely to receive a signal of a frequency either ten percent higher or the same amount lower.

For example, if the signal is at 30,000 cycles a tuned circuit might be expected to discriminate against an interfering signal of 33,000. If the signal is at 300,000 cycles a tuned circuit might discriminate against an interfering signal of 330,000 cycles, but an interference at 303,000 cycles would be very 247troublesome indeed. It couldn't be "tuned out" at all.

Now suppose that the desired signal is at 300,000 cycles and that there is interference at 303,000 cycles. We provide a local oscillator of 270,000 cycles a second, receive by this "super-heterodyne" method which I have just described, and so obtain an intermediate frequency. In the output of the first detector we have then a current of 300,000–270,000 or 30,000 cycles due to the desired signal and also a current of 303,000–270,000 or 33,000 cycles due to the interference. Both these currents we can supply to another tuned circuit which is tuned for 30,000 cycles a second. It can receive the desired signal but it can discriminate against the interference because now the latter is ten percent "off the tune" of the signal.

You see the question is not one of how far apart two signals are in number of cycles per second. The question always is: How large in percent is the difference between the two frequencies? The matter of separating two effects of different frequencies is a question of the "interval" between the frequencies. To find the interval between two frequencies we divide one by the other. You can see that if the quotient is larger than 1.1 or smaller than 0.9 the frequencies differ by ten percent or more. The higher the frequency the larger the number of cycles which is represented by a given size of interval.

While I am writing of frequency intervals I want to tell you one thing more of importance. You remember 248that in human speech there may enter, and be necessary, any frequency between about 200 and 2000 cycles a second. That we might call the range of the necessary notes in the voice. Whenever we want a good reproduction of the voice we must reproduce all the frequencies in this range.

Suppose we have a radio-current of 100,000 cycles modulated by the frequencies in the voice range. We find in the output of our transmitting set not only a current of 100,000 cycles but currents in two other ranges of frequencies. One of these is above the signal frequency and extends from 100,200 to 102,000 cycles. The other is the same amount below and extends from 98,000 to 99,800 cycles. We say there is an upper and a lower "band of frequencies."

All these currents are in the complex wave which comes from the radio-transmitter. For this statement you will have to take my word until you can handle the form of mathematics known as "trigonometry." When we receive at the distant station we receive not only currents of the signal frequency but also currents whose frequencies lie in these "side-bands."

No matter what radio-frequency we may use we must transmit and receive side-bands of this range if we use the apparatus I have described in the past letters. You can see what that means. Suppose we transmit at a radio-frequency of 50,000 cycles and modulate that with speech. We shall really need all the range from 48,000 cycles to 52,000 cycles for one telephone message. On the other hand if we 249modulated a 500,000 cycle wave by speech the side-bands are from 498,000 to 499,800 and 500,200 to 502,000 cycles. If we transmit at 50,000 cycles, that is, at 6000 meters, we really need all the range between 5770 meters and 6250 meters, as you can see by the frequencies of the side-bands. At 100,000 cycles we need only the range of wave-lengths between 2940 m. and 3060 m. If the radio-frequency is 500,000 cycles we need a still smaller range of wave-lengths to transmit the necessary side-bands. Then the range is from 598 m. to 603 m.

In the case of the transmission of speech by radio we are interested in having no interference from other signals which are within 2000 cycles of the frequency of our radio-current no matter what their wave-lengths may be. The part of the wave-length range which must be kept clear from interfering signals becomes smaller the higher the frequency which is being modulated.

You can see that very few telephone messages can be sent in the long-wave-length part of the radio range and many more, although not very many after all, in the short wave-length part of the radio range. You can also see why it is desirable to keep amateurs in the short wave-length part of the range where more of them can transmit simultaneously without interfering with each other or with commercial radio stations.

There is another reason, too, for keeping amateurs to the shortest wave-lengths. Transmission of radio signals over short distances is best accomplished by 250short wave-lengths but over long distances by the longer wave-lengths. For trans-oceanic work the very longest wave-lengths are best. The "long-haul" stations, therefore, work in the frequency range immediately above 10,000 cycles a second and transmit with wave lengths of 30,000 m. and shorter.

Pl. XII.–Broadcasting Station of the American Telephone and Telegraph Company on the Roof of the Walker-Lispenard Bldg. in New York City Where the Long-distance Telephone Lines Terminate.

251 LETTER 24
BY WIRE AND BY RADIO

The simplest wire telephone-circuit is formed by a transmitter, a receiver, a battery, and the connecting wire. If two persons are to carry on a conversation each must have this amount of equipment. The apparatus might be arranged as in Fig. 129. This set-up, however, requires four wires between the two stations and you know the telephone company uses only two wires. Let us find the principle upon which its system operates because it is the solution of many different problems including that of wire-to-radio connections.

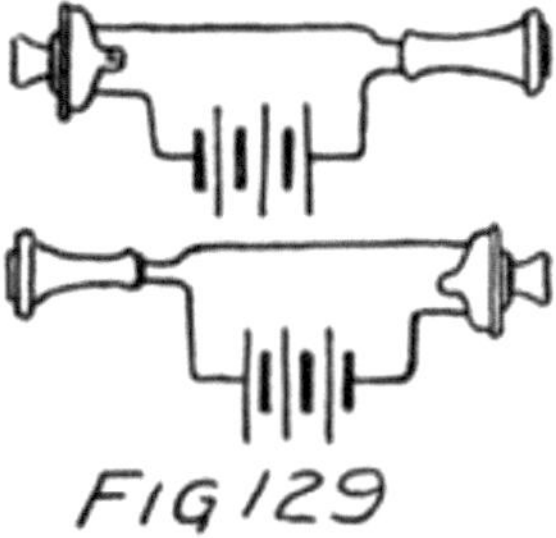

FIG 129

Imagine four wire resistances connected together to form a square as in Fig. 130. Suppose there are two pairs of equal resistances, namely R_1 and R_2 , and Z_1 and Z_2 . If we connect a generator, G, between the junctions a and b there will be two separate streams of electrons, one through the R-side and the other through 252 the Z-side of the circuit. These streams, of course, will not be of the same size for the larger stream will flow through the side which offers the smaller resistance.

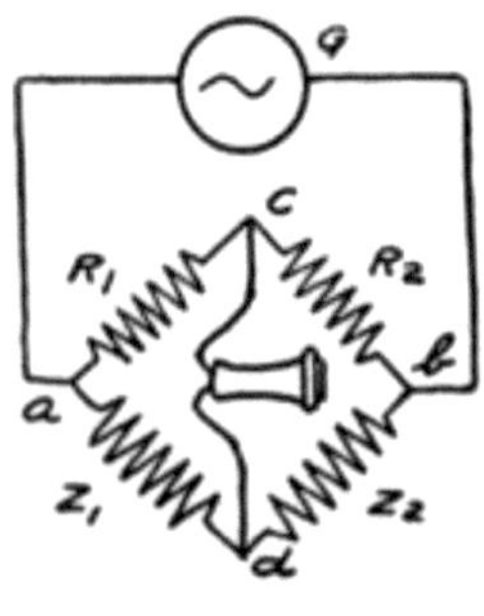

FIG 130

Half the e. m. f. between *a* and *b* is used up in sending the stream half the distance. Half is used between *a* and the points *c* and *d*, and the other half between *c* and *d* and the other end. It doesn't make any difference whether we follow the stream from *a* to *c* or from *a* to *d*, it takes half the e. m. f. to keep this stream going. Points *c* and *d*, therefore, are in the same condition of being "half-way electrically" from *a* to *b*. The result is that there can be no current through any wire which we connect between *c* and *d*.

Suppose, therefore, that we connect a telephone receiver between *c* and *d*. No current flows in it and no sound is emitted by it. Now suppose the resistance of Z_2 is that of a telephone line which stretches from one telephone station to another. Suppose also that Z_1 is a telephone line exactly like Z_2 except that it doesn't go anywhere at all because it is all shut up in a little box. We'll call Z_1 an artificial telephone line. We ought to call it, as little children would say, a "make-believe" telephone line. It doesn't fool us but it does fool the electrons for they can't tell the difference between the real line Z_2 and the artificial line Z_1. We can make a very good artificial line by using a condenser and a resistance. The condenser introduces something of the capacity effects 253 which I told you were always present in a circuit formed by a pair of wires.

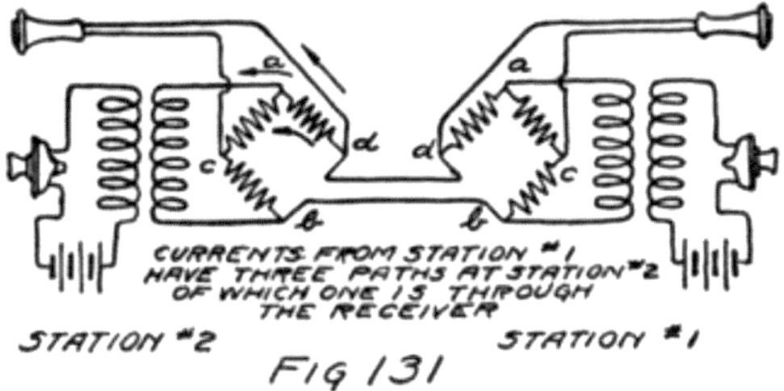

FIG 131

At the other telephone station let us duplicate this apparatus, using the same real line in both cases. Instead of just any generator of an alternating e. m. f. let us use a telephone transmitter. We connect the transmitter through a transformer. The system then looks like that of Fig. 131. When some one talks at station 1 there is no current through his receiver because it is connected to c and d, while the e. m. f. of the transmitter is applied to a and b. The transmitter sets up two electron streams between a and b, and the stream which flows through the Z-side of the square goes out to station 2. At this station the electrons have three paths between d and b. I have marked these by arrows and you see that one of them is through the receiver. The current which is started by the transmitter at station 1 will therefore operate the receiver at station 2 but not at its own station. Of course station 2 can talk to 1 in the same way.

The actual set-up used by the telephone company 254is a little different from that which I have shown because it uses a single common battery at a central office between two subscribers. The general principle, however, is the same.

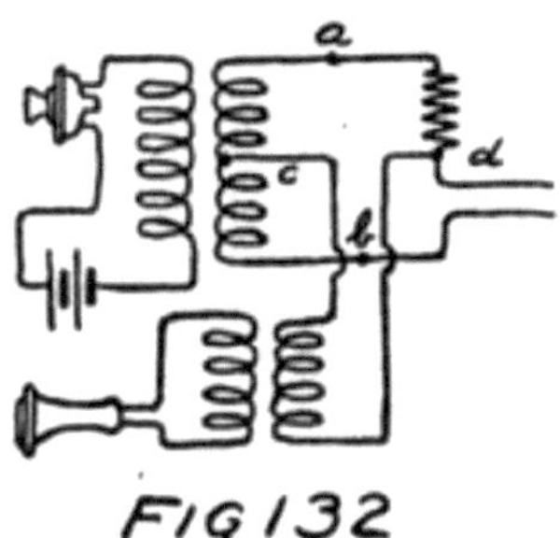

FIG 132

It won't make any difference if we use equal inductance coils, instead of the R-resistances, and connect the transmitter to them inductively as shown in Fig. 132. So far as that is concerned we can also use a transformer between the receiver and the points c and d, as shown in the same figure.

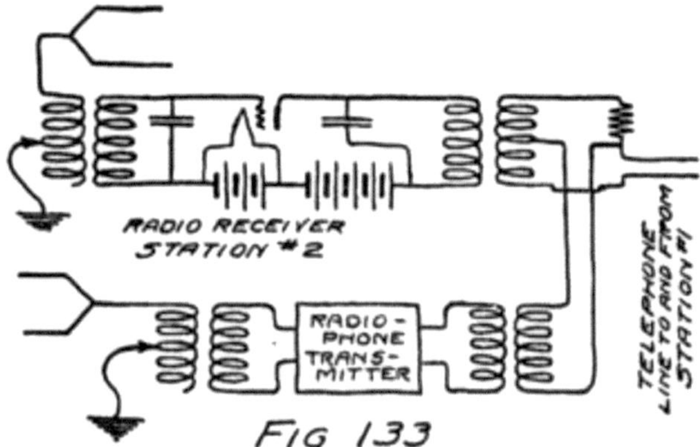

We are now ready to put in radio equipment at station 2. In place of the telephone receiver at station 2 we connect a radio transmitter. Then whatever a person at station 1 says goes by wire to 2 and on out by radio. In place of the telephone transmitter 255at station 2 we connect a radio receiver. Whatever that receives by radio is detected and goes by wire to the listener at station 1. In Fig. 133 I have shown the equipment of station 2. There you have the connections for wire to radio and vice versa.

One of the most interesting developments of recent years is that of "wired wireless" or "carrier-current telephony" over wires. Suppose that instead of broadcasting from the antenna at station 2 we arrange to have its radio transmitter supply current to a wire circuit. We use this same pair of wires for receiving from the distant station. We can do this if we treat the radio transmitter and receiver exactly like the telephone instruments of Fig. 132 and connect them to a square of resistances. One of these resistances is, of course, the line between the stations. I have shown the general arrangement in Fig. 134.

You see what the square of resistances, or "bridge" really does for us. It lets us use a single pair of wires for messages whether they are coming or going. It does that because it lets us connect a transmitter and also a receiver to a single pair of wires in such a way that the transmitter can't affect the receiver. Whatever the transmitter sends out goes along the wires to the distant receiver but doesn't affect the receiver at the sending station. This bridge permits this whether the transmitter and receiver are radio instruments or are the ordinary telephone instruments.

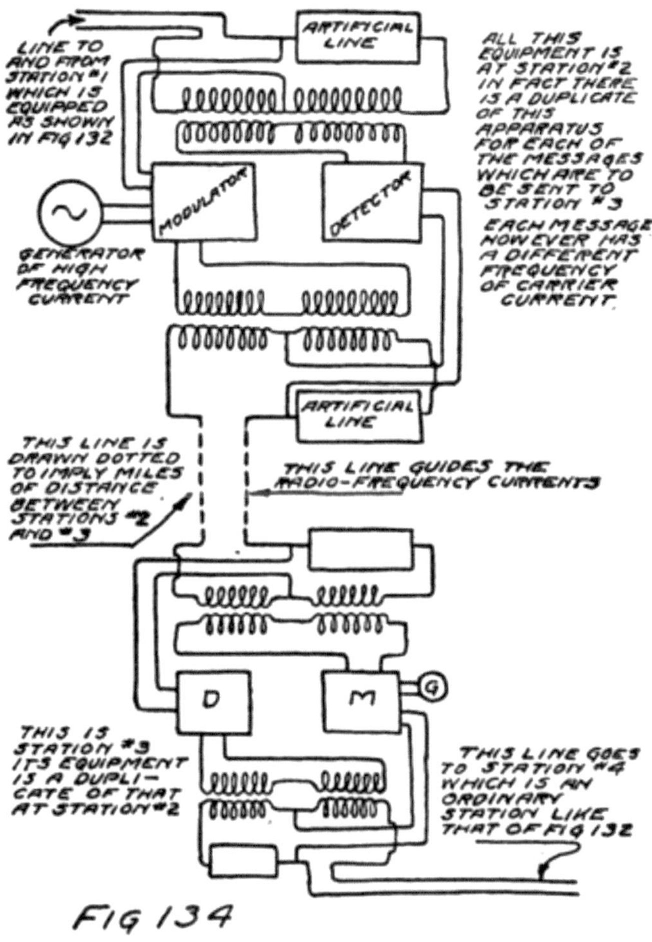

FIG 134

256By its aid we may send a modulated high-frequency current over a pair of wires and receive from the same pair of wires the high-frequency current which is generated and modulated at the distant end of the line. It lets us send and receive over the same pair of wires the same sort of a modulated current as we would supply to an antenna in radio-telephone 257transmitting. It is the same sort of a current but it need not be anywhere near as large because we aren't broadcasting; we are sending directly to the station of the other party to our conversation.

If we duplicate the apparatus we can use the same pair of wires for another telephone conversation without interfering with the first. Of course, we have to use a different frequency of alternating current for each of the two conversations. We can send these two different modulated high-frequency currents over the same pair of wires and separate them by tuning at the distant end just as well as we do in radio. I won't sketch out for you the tuned circuits by which this separation is made. It's enough to give you the idea.

In that way, a single pair of wires can be used for transmitting, simultaneously and without any interference, several different telephone conversations. It takes very much less power than would radio transmission and the conversations are secret. The ordinary telephone conversation can go on at the same time without any interference with those which are being carried by the modulations in high-frequency currents. A total of five conversations over the same pair of wires is the present practice.

This method is used between many of the large cities of the U. S. because it lets one pair of wires do the work of five. That means a saving, for copper wire costs money. Of course, all the special apparatus also costs money. You can see, therefore, that 258this method wouldn't be economical between cities very close together because all that is saved by not having to buy so much wire is spent in building special apparatus and in taking care of it afterwards. For long lines, however, by not having to buy five times as much wire, the Bell Company saves more than it costs to build and maintain the extra special apparatus.

I implied a moment ago why this system is called a "carrier-current" system; it is because "the high-frequency currents carry in their modulations the speech significance." Sometimes it is called a system of "multiplex" telephony because it permits more than one message at a time.

This same general principle is also applied to the making of a multiplex system of telegraphy. In the multiplex telephone system we pictured transmitting and receiving sets very much like radio-telephone sets. If instead of transmitting speech each transmitter was operated as a C-W transmitter then it would transmit telegraph messages. In the same frequency range there can be more telegraph systems operated simultaneously without interfering with each other, for you remember how many cycles each radio-telephone message requires. For that reason the multiplex telegraph system which operates by carrier-currents permits as many as ten different telegraph messages simultaneously.

You remember that I told you how capacity effects rob the distant end of a pair of wires of the alternating current which is being sent to them. That is 259always true but the effect is not very great unless the frequency of the alternating current is high. It's enough, however, so that every few hundred miles it is necessary to connect into the circuit an audion amplifier. This is true of carrier currents especially, but also true of the voice-frequency currents of ordinary telephony. The latter, however, are not weakened, that is, "attenuated," as much and consequently do not need to be amplified as much to give good intelligibility at the distant receiver.

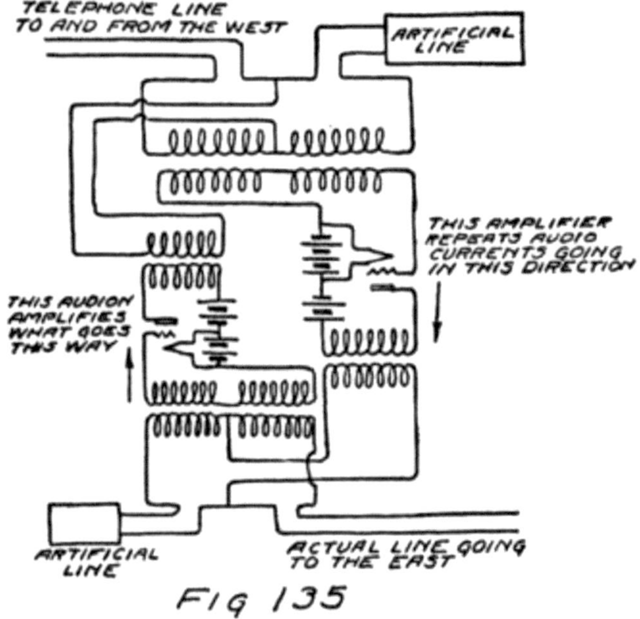

FIG 135

In a telephone circuit over such a long distance as from New York City to San Francisco it is usual to insert amplifiers at about a dozen points along the route. Of course, these amplifiers must work for transmission in either direction, amplifying speech on its way to San Francisco or in the opposite 260direction. At each of the amplifying stations, or "repeater stations," as they are usually called, two vacuum tube amplifiers are used, one for each direction. To connect these with the line so that each may work in the right direction there are used two of the bridges or resistance squares. You can see from the sketch of Fig. 135 how an alternating current from the east will be amplified and sent on to the west, or vice versa.

244

261There are a large number of such repeater stations in the United States along the important telephone routes. In Fig. 136 I am showing you the location of those along the route of the famous "transcontinental telephone-circuit." This shows also a radio-telephone connection between the coast of California and Catalina Island. Conversations have been held between this island and a ship in the Atlantic Ocean, as shown in the sketch. The conversation was made possible by the use of the vacuum tube and the bridge circuit. Part of the way it was by wire and part by radio. Wire and radio tie nicely together because both operate on the same general principles and use much of the same apparatus.